RECALL

A TRUE STORY OF LOVE, LIFE, AND HONOR

Lt. Col. "Art" Walker, Ret.

Foreword by
Terry J. Walker, MA

A Gap Closer™ Publication
Life On Purpose Publishing
An Angela Massey Imprint

SAN ANTONIO, TEXAS

Copyright © 2018 by Lt. Col. "Art" Walker, Retired.

All rights reserved. No part of this publication may be reproduced, distributed or transmitted in any form or by any means, including photocopying, recording, or other electronic or mechanical methods, without the prior written permission of the publisher, except in the case of brief quotations embodied in critical reviews and certain other noncommercial uses permitted by copyright law. For permission requests, contact: terry@iamterryjwalker.com or visit the website: www.iamterryjwalker.com.

Book Layout ©2018 by Dr. Angela D. Massey

Ordering Information:

Special discounts are available on quantity purchases by corporations, associations, and others. For details, contact the Terry J. Walker at the email address or website above.

Cover: pixelstudios
Book Layout: Dr. Angela D. Massey
Reprint: Life On Purpose Publishing

Recall/ Lt. Col Arthur R. Walker. —2nd printing.

ISBN: 978-0-9961908-8-6

Library of Congress Control Number: 2018935834

Contents

Dedication vii

Foreword ix

The Beginning 1

Enlistment 9

Entering the Army 11

Basic Training 15

Advance Unit Training 23

Shipping Out 33

Oran North Africa 39

Combat 45

Cape Bon German Evacuation 49

Replacement Depot 55

Airborne and Sicily 61

Naples, Italy 65

S.I.A.M. Company 71

Florence, Italy 75

Poe Valley 81

End of the War in Italy 85

Civilian Life: 1946-1952 91

Recalled to Active Army Duty 99

First Command 103

Korea 109

Chitose Private Rental 117

Move to Honshu 125

Change of Command 133

Return to the U.S. 139

Infantry School 145

Return to Korea 151

Professor of Military Science 161

Green Berets 165

Omaha 177

Washington, DC 183

Germany 187

France 191

Fort Riley 195

Vietnam 201

Thailand 213

Thai Operations in Nam 219

Fort Campbell 225
Hospitalization and Recuperation 231
Final Review 239
Conclusion 245
About The Author 249

Dedication

First and foremost, I dedicate this book to the stabilizing influence of my wonderful, loving wife, Lois Jean, who laughed or cried with me through the good times and bad times. She willingly accepted the many hardships I caused her.

I leave this book as a legacy to my two children, Arthur Jr. and Teresa Jean.

I wish to say to all my military comrades, living and dead, you will always be remembered in my mind. One day we will again be reunited in another world of "eternal peace."

Foreword

*T*HE STORY you are about to read is by my father, Lt. Col. Arthur R. Walker, US Army Ret.

It is a true story of love, life and honor for his family and for his country as he fought in and survived three wars: WWII, Korea, and Vietnam.

As you read his story, you will get a sense of the conditions he lived and survived in. You may even be angered when you read about the military's lack of support for the soldiers' basic needs. You will also get a first-hand look at the obstacles he faced when he returned home to his family. When my father originally wrote this book in the 1990's, there was no Internet; he published it on his own, and many historians interviewed him. However, the only books that he had personally printed were the only ones left in existence today, and since I have been asked by many to read it and to have access to it, I am republishing the book to allow for anyone to learn from and gain a sense of our wonderful nation's history.

As I was asked to write the Foreword, I could not do so without mentioning the evolution of Post-Traumatic Stress Disorder, (PTSD). Derived from the U.S. Department of Veterans Affairs — PTSD: National Center for PTSD

Evolution of PTWWII Combat Stress Reaction (CSR) also known as "battle fatigue" was a descriptive that the soldiers were just weary and exhausted. There were even some military leaders, including General George S. Patton, who refused to believe that battle fatigue was real. After WWII, the military classified most discharges as just "combat exhaustion."

In 1952, the American Psychiatric Association came out with a diagnosis of GSR-Gross Stress Reaction. The APA believed that those in combat who had experienced traumatic events would resolve themselves relatively quickly after returning home from battle.

It wasn't until 1980 that Post Traumatic Stress Disorder, PTSD was added to the DSM III, (Diagnostic Statistical Manual). It was added primarily because of research involving returning Vietnam War Veterans, Holocaust survivors, sexual trauma victims, etc.

As you can see, there was no accurate diagnosis nor any treatment for soldiers returning home from WWII, Korea, and Vietnam. The military assumed they would be fine upon returning to the States and their families and they could just go on with life as though nothing had happened or changed.

These brave men and women endured, lived in and experienced things that most of us will never be able to comprehend. They put their lives on the line in battle so that we can experience the freedoms we have today. Because these brave men and women fought for us, we have freedom of speech; the right to vote for a political office and the choice to attend college. We can worship whenever and wherever we choose and have a career of our choosing.

As I stated above, there was no Post Traumatic Stress Disorder diagnosis and no type of treatment to help my dad through any readjustment period after a battle. You will read where he was blown off an armored personnel carrier when it hit a land mine, the extent of casualties, along with his injuries. You will also read that upon being released from Walter Reed Hospital, the psychiatrist's report stated he suffered from extreme "combat stress syndrome" (CSR). My father asked, "What does that mean?" and the psychiatrist stated, "That's the nicest way I know to tell you that you're crazy."

One of my earliest memories of my father was after he was sent home from Walter Reed. I was around four or five years old. He came home in a full body cast with only a circle cut out of the cast for his face and a couple of holes cut out for his ears. At first, I was terrified of him. During this time there was a television show called *The Munsters* of which I loved to watch. So, in an effort for me to be more comfortable around him and physically go into the room where he was lying in bed, my mom attempted to persuade me into believing he was now part of *The Munsters* TV show. It worked! I even convinced my friends to come over and see him, and we would draw on his body cast. As the years went by, my "Pop" as I affectionately referred to him, always enjoyed telling that story.

Keep in mind that at that time there was no intensive nor formal rehabilitation as would be available today. My dad came home in a full body cast, and the doctors predicted that he would never walk again. However, my pop proved them wrong! With my mother's persistence, he endured exercising and walking in the woods to rebuild his strength and to rehab himself. He raised English setters and trained them as bird hunting dogs and would go to the woods, train and work the dogs to hunt. So, to continue to try and walk, he had to endure the pain and walking and do rehab on his own. He was never in a wheelchair. Even in his elderly years, while struggling at times to walk, he refused to use a cane.

After many years of personal torment, whether physical or psychological, my mother was finally able to get my father to open up and write about his experiences to work through his haunting past. As you will see, my dad dedicated this book to my mother as she too endured the hardships right alongside of him. He wrote the book originally as an effort to release some of the pain and memories he had experienced. However, as the book transpired, it became so much more. He entitled the book *Recall* as appropriate of his memories and his experiences of all that he endured in those three wars.

Pop left us in 2017 at the age of 93. He was one of the last remaining most decorated veterans. I have had the book republished so it can be accessible to anyone who would love to understand our history from one who lived it.

I hope you enjoy *Recall*, and I pray it will help you to gain a better understanding of what our brave military men and women endured over the years so that we can enjoy the freedoms we have and possibly even take for granted today.

Recall is my dad's words and his experiences, just as he wrote them…just as he recalled them.

Terry J. Walker, MA

CHAPTER 1

The Beginning

BORN IN 1923, I was the second son of (at that time) a successful family. My father was instrumental in and grew with the Rural Electrification Program in Indiana. During the first eight years of my life, my father was very prosperous which enabled my brothers, sister, and I to live a very good life. We lived in fine homes and had plenty of the best food, clothes, toys, and pets. My father, now a big-time supervisor in the Public Service Company, invested heavily in the stock market. I remember hearing of one name, a Sam Ensill (something similar to that) who absconded with a great deal of Public Service money in which my father had invested.

Along came the stock market crash of the 30s and with it my father and my life crashed. My father had let success become his downfall. He was named in a paternity suit by his secretary which led to a divorce from my mother. This left my mother with four children to raise with no means of support. My father went from a wealthy paid supervisor to a WPA job, building shit houses for $12 a week. My mother, brothers, sister and I moved in with her father and mother. They lived in a shack in the slums of Indianapolis.

My grandfather who had one leg amputated because of an injury as a young man, managed to get a job as a night watchman at a used car lot where he was paid $8 a week, and this had to support seven people. It had to provide rent, food, clothes, and no luxury items, which we once had plenty of. In order to get enough money for my brothers and I to go to a movie on Saturday, (which cost a nickel a piece) we would pick up scrap after school. I remember we received a penny a pound for scrap iron, three cents a pound for copper or brass, and a nickel a pound for aluminum. The junk man would give us a penny for five pounds of newspaper or rags. There was no public trash pickup at this time, which was a benefit to us boys.

During this time, I was caught stealing a toy from a dime store. I was taken to the juvenile hall, and later turned over to my mother with the provision that she control my behavior or turn me over to my father. I was not easy to control, so I was sent to my father, who now had married his former secretary. They were living with their baby on a 20-acre broken down farm where they had accumulated five or six renegade cows. These six cows became my responsibility and I had to feed, water, and clean up after those damned contrary bovines. I remember one of them, which my dad bought at a sale, jumped out of the truck when they were being unloaded. We chased her for a week before we caught her. She could jump a fence like a deer, which finally led to her undoing. She jumped a fence around the edge of a pit that was about five feet deep. We were able to rope her and tie her to a pick-up truck. She would lie down and bellow, but we drug her home minus much of her hair. Then I had to milk the damned thing.

By now, my father was selling cars and making some money, but not enough to pay alimony to my mother for the support of my brothers and sister. When the courts started to insist he pay support he decided to slip out of the state of Indiana. He and his wife had made a trip to Tennessee where he could buy land for $10 an acre, which had been overpriced to him by $5 an acre, but he managed to secure 120 acres somehow. This land was on top of the Cumberland Mountain in Tennessee, one of the most scenic areas in the United States. It was a rugged country known as the poverty area called Appalachia.

He loaded his wife and baby in a new Chevy pickup truck (not paid for) and hooked to a rubber tired, four-wheeled wagon with a short tongue to be used as a trailer and later to be used as a farm wagon. We piled all his junky furniture, tools, and possessions on this wagon. He even made a cage on the back, in which he put two calves and five pigs. This wagon was my carriage, I rode on the top of everything on an old mattress exposed to the world. I remember on the second day of the trip we had to go up a mountain to the Cumberland Plateau. My job was to carry a big chunk of firewood to throw under the wheels in the event the brakes failed. The brakes didn't fail, so there was no failure on my part, but I was sure tired from trotting almost a mile with that old chunk of wood. I forgot to mention we had 12 stinking chickens on that wagon. Today, we would have looked like gypsies from hell, but during that period of time with so much poverty we were hardly noticed.

Finally, we arrived on the Cumberland Mountains in Tennessee where someone let us live in an old house half of which had fallen down. It was about two miles to the acres my dad had bought, and when he went off in the truck to try to find work for a little cash, I had to walk to the acres carrying a double bit ax. You notice I refer to this place as acres and not a farm. You see this place was covered with burnt over second growth timber, red brush, snakes, and saw briars. We had to clear all this land with axes and a crosscut saw, burn the brush, and grub the stumps, this was before chainsaws and power tools. I was one end of the power on a crosscut saw, and my father was the power on the other end. My father was constantly accusing me of riding the saw on my end. In retrospect, I believe this was an unfair accusation when you realized my father, a stout grown man was on one end, and I a 12-year-old small but wiry boy on the other end.

I worked on these acres a year and started high school in Crossville, Tennessee. I was in a vocational training class, in which we learned and worked at building the school. This was a good experience which helped me in later life and might be a good program to revitalize in today's society. It was a part of the reconstruction program that gave many a young person something to do that was constructive, taught them a trade, and enabled them to build strong bodies. This work made you so tired you didn't need all the fancy recreation programs young people think they need today.

During this period, Crossville, Tennessee held a big area sheep sale. My father got a job there and he used me to help separate the sheep into pens so they could be sold in groups by the owners. For a week's work I was given a black lamb which I sold for $.75. Two days after the sale, my father's wife told my father I had worn my only pair of shoes (a pair of basketball shoes my uncle had given me) to clear the new ground instead of going barefooted. I had worn the shoes to keep from cutting my feet on the brush stubble and the saw briars.

My father was drinking heavily at this time, so he gave me a whipping. I told him then, that was the last whipping I would ever take. That night I left Tennessee and hitchhiked about 500 miles back to Indiana. I was 13 years old, and on the way, I spent part of the $.75 that I had made herding sheep into pens. I bought six hamburgers for a quarter, a Barlow knife for a quarter, and I had a quarter left when I got to my grandparents in Indiana. My plan was to sleep behind a billboard that had a light on it. I was afraid of the dark and I was going to use the Barlow knife to protect myself. As it turned out, the last man who picked me up was going to work on the night shift at a cement mill in Speed, Indiana. He suggested I sleep on the scale pen under the empty cement bags. During this time there were many hobos of all ages on the road, so I was no rarity. The man told me that many drifters slept on the scale pens at night when they were not in use, so that is where I spent the night.

When I arrived at my grandparents in Indiana, my brothers and sisters were living with them. My grandfather was able to get some carpenter work and worked part-time in a rural butchering house, but his finances were meager, and I would have imposed an additional burden. I had a distant relative who managed a big dairy farm who took me in. He had a son my age and an older daughter, and we all worked very hard. Up every morning before 4 a.m. and to bed after the chores were done about 9:00 o'clock p.m. We each had our assigned chores. I milked 20 cows morning and night by hand, as there were no milking machines then. This kind man and his wife treated me as their own. They gave me $.50 a week, bought me some overalls and shoes. They sent me to school with a good lunch and never demanded anything more from me than from their own children.

I want to add here that none of these people, not my grandparents, my mother, my father or any of these relatives ever accepted any public relief. I am not even sure there was such a thing as a relief program, but I am sure there was poverty. The nearest thing to a relief program I can think of was soup lines where people would get a free bowl of soup. The people I knew during this period were survivors, too proud to accept relief. To conclude this chapter in my life, it is sufficient to say I finished school, not claiming to be a very good student, but I did graduate from high school.

I knew I would have to leave this small rural community if I ever intended to find a wife as both my father and mother had grown up there. I was related to nearly all the people, either on my mother's side of the family or my father's, so any liaison I was to make would probably have been incestuous.

The nearest town of any size was 15 miles away and the people I was living with would let their son have their old car to drive to town on Saturday nights. He and I would fill the car with $.13 a gallon of gasoline. Five gallons of gas would get us to the big city where we could ride up and down the streets whistling at the girls. I think this same procedure is done by young people of today, but at a much higher cost. I believe today's term for this action is called "cruising." History has a way of repeating itself generation after generation. The only difference I note in this procedure is the cost and the frequency. It seems young people cruise every night and gasoline is more than a dollar a gallon.

I became a survivor at a very young age. I learned early that I could take care of myself and I did not need to rely on others. I do not criticize my parents or anyone else for the mistakes they made. I developed a philosophy that everyone had to do what was necessary to exist, and each had to answer for their own mistakes. Growing up under these conditions, I learned responsibility and developed self-confidence. I even concluded that any person could accomplish anything if he was willing to put forth the mental and physical effort required.

CHAPTER 2

Enlistment

WHEN I was 17 years of age I decided to enlist in the service. The United States was on the verge of World War II. The song of the day was *"Goodbye Dear, I'll Be Back in a Year."* Young men were being drafted for one year of training only. I was too young for the draft and having no family ties I had to enlist. The war effort was creating many jobs, but there were so many unemployed, the jobs were being filled by older more experienced men. Later during the war when so many men entered the service, women took over the workforce. A teenaged high school graduate with cow milking experience wasn't too much in demand unless he elected to join the military.

I took and passed the mental exams to enlist in the Navy V5 program. I just knew I would be the best Navy fighter pilot in America. I could see myself in a beautiful uniform landing a slick new airplane on the deck of a carrier. That uniform to a boy who had been whipped for wearing his only pair of shoes was a big incentive. With all my high ambition I was sent to the

Great Lakes Naval Station in Chicago with a good friend of mine from high school for a final physical exam. He passed, I failed; they found I was color blind. He was elated, I was disappointed; they even informed me that the regular Navy would not accept me. My friend went on to become a good Navy pilot, was commissioned, and served in the Pacific with distinction.

Next, I tried the Marines. They said I was too small, so there went that beautiful dress uniform with the high collar, which probably was a good thing because I have a very short neck and the collar would have been very uncomfortable. Had my neck been longer I might have been tall enough, but that was the hand fate dealt me.

I tried the Merchant Marines, but I was not too impressed with them because at that time there were lots of news stories about merchant ships being sunk and people floating around in life rafts with the sharks. Besides, they didn't have nice uniforms and I had never been on a ship or seen an ocean.

After all these disappointments there was only one place left for me to go. I joined the Army. They took anyone, or so it seemed at the time. If the Army hadn't accepted me I could have tried the Coast Guard. However, Indianapolis, Indiana didn't have a Coast Guard recruiting station, as there isn't any coast there to guard.

The Army said they would accept me providing I could get my parents written consent. I started asking my mother to sign the necessary papers and at first she would not do it. I told her I would run away and enlist somewhere else where I could lie about my age. She finally gave in and agreed the Army might do me some good.

I was now able to get myself a nice new uniform even though it was O. D. (Olive Drab) and I now had two pairs of brand new brown shoes.

CHAPTER 3

Entering the Army

*M*Y STORY and maybe my life really began here. The reader should realize many of the following events took place more than 50 years ago. Some of the episodes may not have occurred as I describe them; however, they are based on truths or hearsay to the best of my recollection.

I propose to show the conditions as they were in the Army starting in early 1942. I portray some of these events rather graphically so that the public will understand how life really was in the Army at the beginning of World War II. Any persons or units depicted during my World War II experiences, other than my own, will be given nicknames or glossed over to protect their identity. Many soldiers who served during World War II, no doubt, will recognize or recall some of these experiences, especially if they served in the European Theater of Operations.

In order to enlist in the Army I had to get my mother to sign a consent form as I had just turned 17, this she reluctantly did at the recruiting station. At the recruiting station a Sergeant asked me what unit I wanted to join, since I was enlisting I could choose where I wanted to go.

Since I knew nothing about the Army he suggested the Coast Artillery. He said I would be given basic training, then assigned to a unit near a big city somewhere on the U.S. coastline. This seemed like a good deal to me. I would remain in the U. S. firing big guns and going on pass every weekend, dressed in my Army uniform meeting all kinds of pretty city girls. So, I chose to join the Coast Artillery. What they didn't tell me was that the Coast Artillery had just included the mission of Anti-Aircraft Artillery due to the attack on Pearl Harbor. Since there was a great need in this area, that is where I was initially assigned.

My first day in the Army at Fort Harrison, Indiana was filled with shots, drawing uniforms, and being assigned a cot in a barracks. This first day passed quite uneventful except for getting sick and sore from the immunization shots.

The second day I was awakened at 4 a.m. and told I was on KP. I had to ask what KP was, and when I was told it was kitchen police, I had visions of standing guard in a mess hall which I would be proud to do. How wrong I was. For 12 hours, I washed dishes, cleaned garbage cans, swept floors, stoked coal, fired ranges, and listened to a swearing Sergeant telling me what a sorry do-nothing Private I was. I got through that second day still sick and hurting from my shots. I returned to my bunk exhausted, anticipating a good night's sleep. I fell out of my bunk still clothed which turned out to be a big mistake.

About two o'clock in the morning I was rudely awakened by a Corporal who told me I was on fire guard. I guess because I still had my clothes on I was the easiest one to pick, so he explained what a fire guard did. A fire guard walked around the outside of four barracks for two hours, watching to see they did not catch on fire. If they caught fire the guard was to sound the alarm. I remember this was in November and it was cold as hell. I even thought about setting one of the barracks on fire just so I could get warm. After two hours I was relieved and told I could go back to bed, which I happily did. I was allowed to sleep till about 10 o'clock, when this bastard Corporal came in and told me that because I didn't have shipping orders I was to report to the Officers Mess to act as a table waiter. Table waiting was much easier duty than KP, which I learned to use to my advantage later in my Army service. As a table waiter, all I had to do was fill the bowls with food as the officers emptied them. After about six hours of waiting on officers to whom I was told I should not speak, I returned to my bunk. I went to bed and slept like a log. This time I removed my clothes and slept under my blankets. I knew I didn't want to be a fire guard again.

It was now Saturday and in came the Corporal. This time he was a little more friendly, and he said he would let me go on pass to Indianapolis on Sunday. Before giving me the pass, he explained the penalties for desertion and failing to report back on time. The Corporal said the penalty for desertion was death by firing squad. I called my mother to come pick me up, which she did and took me to her house where I showed off my nice uniform. She acted very proud of me so I never told her that I thought I might have made a big mistake enlisting. We had a great meal and I didn't have to do KP, but I did help wash the dishes. After the lesson on desertion, I reluctantly returned to Fort Harrison where I read the bulletin board, at which time I learned that I was to ship out the next day to Fort Eustis, Virginia to start basic training.

I was still sore and sick from all the immunization shots. I was scared as hell and I was wondering what I had gotten myself into. I was ready to get out of Fort Harrison anyway

I could. If I hadn't been told about being shot by a firing squad for desertion, there is no telling what I would have done.

Seeing my orders on the bulletin board was a relief since I knew what it was like at Fort Harrison, but it also increased my apprehension as I would now be facing another unknown situation. Later in life I came to the realization that this is what occurs in life, going from one known situation to an unknown situation. I started to accept the unknown future as a new adventure with problems to overcome and accept life as a challenge.

CHAPTER 4

Basic Training

I EXPERIENCED my first train ride which was about 12 hours from Indiana to Virginia in a dirty, sooty coach with two barracks bags crowding me in my seat. I was given a sack lunch with a spam sandwich, some cookies, and an apple; water was available from the fountain on the train. It was so happy to be leaving Fort Harrison that I kind of enjoyed the trip.

This was an all troop train that went directly into the post, where we off loaded and were herded like a bunch of cattle onto trucks as we had not yet learned to march as soldiers in a military formation. We were transported to a quadrangle of barracks, assigned bunks, and started orientation proceedings.

Basic training consisted of eight weeks training. I learned to march, make beds, scrub floors, wash windows, and do the everlasting KP. I was taught to say "Sir" to everyone and never violate the chain of command, which meant I had to ask a Private First Class for permission to speak to a Corporal and a Corporal for permission to speak to a Sergeant. I never spoke to anyone higher in rank than a Sergeant. I only went to the orderly room if I was summoned, and when there I stood outside with my hat in my hand until I was told to enter. I never got to speak to an officer during my entire basic training, which was probably to my benefit because I did not know how to address such an elite person. All your thinking was done for you. Your individuality was taken from you and you became a robot manipulated by your superiors. That was the purpose of basic training. It was intended to make each man react to orders without question.

I will digress here for a moment to relate how unprepared the Army was in 1942. Our equipment consisted of wrap-around leggings, steel helmets with rims, long bayonets, and gas masks, all of which were left over from World War I.

There were some 1903 model Springfield bolt action rifles available, which were kept under lock and key when not in use. One of these rifles was issued to each nine-man squad and it was rotated among the squad for each man to disassemble, clean, and reassemble. We were never permitted to fire these rifles. It seems there was a terrible shortage of small arms ammunition and the supplies that were available were given to infantry trainees.

For drill and training we were issued mock rifles made from broomsticks nailed to a piece of wood for a stock. We would run, fall, point these wooden rifles and yell, "Bang-Bang." I had played games like this as a little kid; however, now I was doing it as a grown-up soldier.

For a mortar we were given a joint of stove pipe and blocks of wood for ammunition. We would drop this wooden ammunition in the stove pipe, then yell, "Bang," not "Bang-Bang" since the simulated a single fire weapon.

Recall

In Uniform for Guard Duty, 1942

    A piece of stove pipe became a very valuable item. It could be placed in a backpack with a blanket rolled around it. This made a very neat pack, which was much lighter than one made with the heavier items it was supposed to contain. I am positive many simulated mortars went on long marches.

    During basic training I learned to make my cot with square hospital corners. We had to turn our sheet out with an exact 12-inch fold, called a collar. The "U.S." which was stamped on all blankets, had to be in the exact center of the cot. The pillow had to be covered with your second blanket. The cots, the pillows, the collars, and the U.S. stamps then had to be in direct alignment from one end of the barracks to the other. We used a long string to line these items precisely. Everything had to conform to a standard routine that had been created in a peacetime Army.

Every morning the barracks were inspected and if anyone's equipment was out of order you would be a gigged (given demerits). Enough demerits would result in extra duty such as KP. On Saturday morning there was a big inspection which was conducted while standing by our cots. This inspection was conducted by an officer and if you failed you would not be given a pass or Sunday off. To prepare for this Saturday inspection, we would spend all of Friday night scrubbing the floors, washing the windows, and preparing our gear for inspection. When our cots were made we would sleep on the floor under them for what little sleep we were able to get so the cots would be precise enough to pass the inspection.

For all of this work we were paid $21 a month whether we deserved it or not. We were paid once a month at which time we had to donate to the company fund. The pay officer would then collect for our canteen checks and make us pay for any lost or broken equipment. When everything was taken out of our pay we were lucky to receive five dollars. That wasn't all bad because we had our clothes, a place to sleep and plenty to eat (such as it was).

Personal hygiene was taught and rightfully so. Many of our new soldiers had never experienced running water or inside toilets, especially if they had grown up in rural areas. Toilets and showers were called latrines in the Army. To be given the job of latrine orderly was one of the most revolting assignments you could be given. The latrine orderly had to scrub the toilets, sinks, and showers all day. The job was usually given as a form of punishment.

Mass punishment was meted out to a whole platoon or even a complete barracks. If one man failed to wash, shave, or bathe, the whole squad would be punished. I shaved every day even though I wasn't old enough to have a beard. Initially, our heads were shaved, and later allowed to grow to an inch in length. This was to help us keep clean and remove our feeling of individuality. If a soldier failed to wash or shower, the Sergeant would have his squad give him a GI (government issue) bath. A GI bath consisted of the squad undressing the soldier and holding him under a shower where he was scrubbed with GI brushes and GI soap (lye soap). One bath of this nature would make one want to wash by himself. I never received a GI bath but must admit I helped give a few. Another reason other than cleanliness for the short hair was in the event a soldier received a head wound it would aid in preventing infection. To this day, I still wear my hair short even though I am unlikely to receive a head wound unless my wife of 47 years reads this book.

We were given a reason for nearly everything we did, hence the saying, "There's a right way, a wrong way, and the Army way." We had to do everything without question the "Army way." Later in my career I tried to institute another form called, "my way." When I invoked this way it usually ended up with my suffering some type of disaster.

During basic training we were restricted to the base area for the first four weeks. If at the end of the four weeks we hadn't compiled too many gigs (points for rule infractions) we were permitted an overnight pass off post. During the first four weeks I didn't know enough to get gigged, and I had become friendly with some of the training corps so I was allowed a pass.

The training cadre consisted mostly of regular Army enlisted peacetime soldiers. Some had 15 to 20 years of service.

Many were alcoholics, some with criminal records, some were derelicts, most unmarried and many had found a place to escape from society. They had a place to sleep, clothes to wear, and three meals a day. Remember, World War II was the beginning of the end of the Great Depression, and I was told that it was very difficult to enlist in the peacetime Army. Many men considered the Army a safe haven from the depression where the necessities of life were furnished with a minimum of responsibility.

My reason for acquainting you with this cadre was to introduce you to one we called "Little Willie." He was a professional Private First Class. with over 10 years of service. He was what we called a "pusher." He woke us up, he pushed us to every formation, he put us on detail, and he yelled at us all day long. He pushed us to hurry in everything we did: rush, rush, rush! He also had the final say as to who went on pass.

Little Willie was an alcoholic who never wanted to be anything more than a Private First Class. We had to keep our foot lockers locked at all times to keep him from drinking our shaving lotion. I was told that he would take sterno disks, melt them, pour them through bread, then squeeze the alcohol out and drink it. Willie was an uneducated society dropout who had found a home in the Army.

Back to my first overnight pass: Little Willie was the man in charge. It was right after payday and we had been trucked into Newport News. We got out of the truck and Little Willie told me to wait while he went into a store to buy a little bottle of olive oil to line his stomach, so he could drink more cheap bootlegged whiskey.

Newport News at that time was considered a sailor's town. I guess it still is, but then, there was a great amount of competition between the services. While waiting for Little Willie, I was standing on the curb when three drunk sailors walking arm in arm came staggering down the street. When they got beside me one of them hauled off and slugged me, knocking me down. They never said a word, they just hit me; well, I got up and lit into one of them. About this time, Little Willie came out of the store and got into the fight, more soldiers arrived, and more sailors came. It wound up as a riot stopping traffic, until someone called the Shore Patrol and the Military Police. When the police got everyone separated, all the soldiers and I were put back on the truck and was shuttled back to Fort Eustis. I don't know what was done about the sailors, I just know that my first 24-hour pass lasted all of 30 minutes.

The rest of my basic training was rather uneventful. We learned a lot of "gold-bricking" habits, like falling out for reveille in the dark wearing only shoes and an overcoat. This would allow us another 10 minutes of sleep time. One morning we got caught when a Sergeant came around with a flashlight. The whole barracks had to double time 20 laps around the quadrangle in unlaced shoes and long, heavy, flapping, brown GI overcoats. This occurred in December and it was cold as hell, but we soon warmed up after doing 20 laps.

In the early 40s the Army uniforms (left over from World War I) were made of brown dyed wool, known as "the old brown shoe Army." Black shoes and green uniforms came much later. The overcoat was long and heavy with brass buttons and a big collar. It covered a soldier from his shoulders to the tops of his shoes. It weighed 8 to 10 pounds when dry, but when it was wet from rain or snow, it doubled in weight. Yet it was warm and protected your body from the elements. However, worn over bare skin as we did that morning during the run, it became a major irritant to your skin. This might seem funny to some people, but it drove home a lesson to us and we always dressed for reveille from that morning on even though it cost us a few minutes of extra sleep.

CHAPTER 5

Advance Unit Training

I FINISHED basic training and embarked on my second train ride, destiny Camp Stewart, Georgia in the heart of the Okefenokee Swamp. I know the government sent scouts to find the most undesirable locations in the world to put Army training camps. I recently visited Fort Stewart and it is now a huge established base with modern buildings, fine facilities, a hospital, golf course, commissary, post exchange, and service clubs.

When I was there for basic training there was one small permanent wooden building used as a camp headquarters. We were billeted in six-man squad tents with dirt floors and a sandy area where the tall black dirty Georgia pines had been cut and removed. I was assigned to a newly formed anti-aircraft battery, which was part of what was called a separate anti-aircraft Battalion. The battalion was made up under the rectangular concept, consisting of four-gun batteries and a headquarters battery. The gun batteries consisted of four-gun squads, each with a 40-millimeter Bofors antiaircraft gun. This was a light, mobile, automatic weapon that proved to be very effective against low-flying aircraft. Each gun squad was equipped with a fifty-caliber water cooled machine gun. This machine gun became my assignment. I was responsible for cleaning it, digging a protective position for its placement, and occasionally I was given permission to fire it.

About the only thing I can recall about advanced unit training was the conditions at Camp Stewart. It was sandy, swampy, full of rattlesnakes, water moccasins, and alligators. We learned to avoid most of these perils when we went on marches through the swamps, but I never was able to overcome my fear of snakes.

There was a tar paper shack across a dirt road from our tent area that was called a PX (post exchange). You could buy a pack of cigarettes for five cents, two candy bars for five cents, or a bottle of 3.2% alcohol beer for ten cents. This beer became known as "horse piss" to differentiate it from whiskey that was called "panther piss." I bought a Bulova wristwatch for $15, the first watch I ever owned even though I saved for three months to pay for it. Everything in the PX at that time was very cheap as it was run by soldiers for soldiers. Later the PX system was commercialized resulting in no savings to the GIs today. Today most of the items stocked in the PX system can be purchased cheaper or as cheap at civilian discount stores.

Outside the PX was a fenced area with picnic tables where the soldiers would drink beer until 10 o'clock. At 9:55 they flicked the lights off and on and if the soldiers had not cleared the area by 10 o'clock, the guards armed with baseball bats cleared the area. I saw many soldiers being drug back to their area with headaches, not necessarily from the beer. I did not drink beer so that was two headaches I was able to avoid, one from the beer and one from a baseball bat.

At this time the pay for a Private went from $21 a month to $50 a month. I thought I would have *so much* money I wouldn't know what to do with all of it. It only took one overnight pass to Savannah to show me what to do with it. I slept in the finest hotel in Savannah. Of course, I slept in a chair in the lobby, because at that time they would let our gallant service men do that. This is where the saying originated, "Nothing is too good for our service men, so we will give them nothing." In all fairness, servicemen were beginning to get encouragement and a degree of respect. Prior to the 40s I was told there was signs that said soldiers and dogs stay off the grass, although I never saw any of those signs.

I recall that at this time the number of soldiers in the armed forces was increasing at a tremendous rate. All those *"Goodbye Dear, I'll Be Back in a Year"* men were retained on active duty instead of being released after their year of training. They, as well as myself, were now in the service for the duration and six months. My enlistment term was to have been three years. I used to think, "My God, I'm in this Army for the duration and six months and hell there is no duration in sight." I was not happy about this situation, but I was more afraid of letting my fellow soldiers down than I was of being shot by a firing squad. I had now been promoted to Corporal and was given a sense of responsibility for others as well as myself.

Since I have grown much older I think about how much better living conditions are for prisoners in a penitentiary in the 80s and 90s than they were for servicemen in the 40s. They serving time for crime, we serving our country. Today, in some prisons the pay is greater than it was for the servicemen then. Today, prisoners have clean cells, beds, toilets, TVs, radios, and they get their mail regularly. None of these amenities were available to the soldiers of the 40s.

One of the worst assignments a doctor could get in the Army was what GIs referred to as a "pecker checker." Each month at an unannounced time, the battery would have to fall out naked except for an unbuttoned raincoat and appear before a doctor for a "short arm" inspection. "Skin it back, milk it down— next." The doctor would shine a flashlight on your pubic hair to see if you had the crabs. This procedure was to detect any appearance of a venereal disease. A venereal disease at that time could be the basis for a court-martial which could result in loss of pay for a month.

Why it was done every month is still a mystery to me because I seldom got out of Camp Stewart more than once every two months. Anyhow, this was the Army way. If you had caught the crabs or "galloping lice" as we called them, the remedy was to shave your pubic hair and paint for private parts with blue ointment." I put the "parts" in to distinguish between the rank of "private" and the private parts of the body.

Anyway, at this time I had been promoted to Corporal, so I don't much care how you interpret this. These whole proceedings were very degrading, but I suppose necessary. A remedy the GIs joked about the best way to get rid of the crabs was to rub alcohol and sand into your pubic hair, then the crabs would get drunk and stone each other to death. It was said if you turned a crap over it did no good because they had legs on both sides.

We went on maneuvers along the coast of Georgia. We strung barbed wire on the beaches, dug foxholes, and fired all our weapons. I swear that in my 22 years of service, I have dug up at least half the earth in the world digging foxholes and latrine trenches. Maneuvers were rather uneventful until the end. All we did was dig, wait, and string barbed wire which I did pretty well having built barbed wire fences on the farms where I had grown up.

The Battalion Commander, a Lieutenant Colonel, was from the Savannah area and he was able to get the local people to open a resort building on Saint Simons Island and sponsor a dance for us. All the resorts on the coast were closed at this time in fear of an invasion. The local political powers were able to induce several hundred "Southern Belles" from the surrounding area to come out and dance with our "fighting boys." A Southern Belle is a term used in the South to describe a young beautiful single lady.

I had never danced before that night, but after sampling a quantity of punch that had been heavily spiked with alcohol, I thought I was the greatest dancer in all the world. I danced with every girl I saw, but one very pretty girl caught my attention and I monopolized her most of the night. She didn't seem to mind, so we were able to get acquainted. I stepped on her feet so many times she asked me to go out on the balcony we could sit down.

While sitting out on the balcony she asked me if I would like to visit her home for a weekend. She lived in Brunswick, Georgia which was about 40 miles from Camp Stewart. I told her I had no means of transportation. She said her father had given her a car for her 18th birthday and she would be happy to pick me up and drive me to her house if I could get a weekend pass. She gave me her phone number and asked me to call her when I could get a pass.

The next morning my unit returned to Camp Stewart, when we unloaded our equipment everyone had to fall out for a short arm inspection.

After the maneuvers we remained at Camp Stewart for nearly a month of intensive training which allowed me two weekend passes. I was able to call the girl I had met at the dance. I asked her if she remembered me, right away she replied she did and wanted to know if I was able to get a pass. When I told her I could, she asked what time I wanted her to meet me. I told her I could meet her at noon on Saturday at the front gate to the camp. She replied she would be there. I had called on Wednesday and I thought Saturday would never come.

Saturday finally arrived. I walked to the front gate expecting to be disappointed, believing she would not really be there. To my elation she was there waiting, she seemed very happy to see me and said she had been at the gate nearly an hour wondering if I would come.

She drove me to her home in the suburbs of Brunswick, where I was in for a big surprise. Her home was a mansion complete with servants. Her father was president of a ship building company and I was getting a little scared at the prospect of meeting him, surrounded by all this elegance.

Her father and mother greeted me very warmly when we entered the house. They introduced themselves and tried to put me at ease. I was shown to a beautiful bedroom and told when I had freshened up they would be having lunch with me. It didn't take me long to wash my face and hands because I was getting hungry.

When I entered the dining area this 19-year-old girl was sitting at the table holding a year-old baby. I asked her whose baby she was holding. When she told me the baby was hers, I really must have looked surprised and said something to the effect that I didn't think she was married. She told me she was not married, and then told me that she had eloped with a boy she had known all her life and when his parents found out about it they were afraid it would interfere with his going to college and preparing himself to take over their business.

Both their parents agreed to have the wedding annulled, although at the time they did not know she was pregnant.

Other than that big surprise about the baby, I spent a wonderful day and night with them. This girl drove me back to the camp on Sunday afternoon, accompanied by her baby.

I was able to spend one more weekend with her and her family. I promised to write her when I left Camp Stewart, which I did most of the time while I was overseas. She was really a very nice person and her letters did much for my morale, but I was never to see her again.

The rest of the time at Camp Stewart, we were restricted to the post for security reasons. We spent the time cleaning, greasing, and packing our equipment. Rumors were flying that we were going somewhere, maybe the Pacific, maybe England, maybe Alaska, maybe California. Who knew? I personally did not care since I had been promoted to Corporal.

Before I leave this segment of my Army life, I wish to pay tribute to some of the best regular Army non-commissioned officers I ever knew. Our First Sergeant, (I wish I could remember his name) was a very mature, understanding, individual that provided great help to the new breed of draftees and enlistees. He was the barrier between commissioned officers and the enlisted people. He performed his duty with great agility which enhanced the harmony in the unit.

The gun platoon was led by a Staff Sergeant whose name was Moore. He was from somewhere in the South, and a better soldier I have never known. He was a *"Goodbye Dear"* who had finished a year of training and was held for the duration. Moore was a very sharp looking person who could have been a movie model for the "perfect soldier."

The machine gun section leader was a Buck Sergeant whom I will call Johnson. He was a full-blooded American Indian and I know he must have had the strongest jaws and teeth in the world. He could put a quarter between his teeth and with his hand he could bend the quarter. Nearly every man in the outfit had a bent quarter. Johnson was a regular Army soldier with over 10 years of service. Every payday the Battery Commander would reduce Johnson to Private, because he would leave to get drunk and wouldn't come back until his money was gone and he had sobered up. But when he did come back to the unit they would promote him back to Buck Sergeant. He was so good with a machine gun the Battery Commander could do little else. During this time in the Army, if you were "busted" you were demoted all the way back to Private, but they could promote you as many grades as they wanted at one time.

The change to the Unified Code of Military Justice Act had not yet been enacted. Civilians had no jurisdiction over service people. If a service person was apprehended by civilian police he had to be turned over to the Military for punishment under the Articles of War. There were three categories of court-martial: Summary, Special, and General. Only in a General court-martial did a defendant get any formal legal counsel. A Summary court-martial was regularly held by a Company or Battery Commander who was usually a Captain. Punishment consisted of reduction in rank, a fine, and restrictions, or you could be put on some dirty detail for an unlimited period of time. A Special court-martial was administered by a Battalion Commander, usually a Lieutenant Colonel, here a soldier could be sent to a stockade for six months as a maximum punishment. I knew little about a General court-martial, except I knew it could give the death penalty; however, I never heard of the death penalty having been given. Soldiers were regularly told that desertion would get them a firing squad.

Discipline was one of the main things taught in the Army during this time. Soldiers were taught to obey all orders from their superior without question. We were regularly given lectures on the penalties for disobeying an order to imprint on our minds what could happen if we failed to do our duty. I recall an adage that went like this: "It's not yours to reason why, but to do or die." This seemed to be the military motto of the time.

To gain unity during advanced training we were fed in mess halls on china plates and bowls family-style. Each squad was seated at a picnic type table with benches. The man seated nearest the center aisle was designated table waiter. It was the duty of the table waiter to refill the big bowls placed in the center of the table when they became empty. If a soldier wanted more food he would ask for it by saying, "Please pass the beans" or whatever it was he wanted, and the bowl was passed to him. If someone else took a helping before it was passed to the man who asked for it, this was a no-no called "short stopping." I saw one man bodily thrown from the mess hall for short stopping and he was not allowed to reenter the mess except to do KP for a week. KP was used as a form of discipline and you could be put on KP scrubbing garbage cans or any other dirty detail, such as latrine orderly, for as long as the Commander wanted. Usually this punishment was accomplished on your free time before or after training hours.

I soon learned to follow orders as did the other soldiers, but I must say these duties never hurt me and later this discipline may have saved my life many times. The methods may have been crude and cruel, but that they were effective. A well-trained, disciplined soldier reacts automatically in critical emergency situations as I did later in combat.

Unit training is very important, and every man must instinctively be able to rely on every other man. Every person must forget his individuality and become a part of a team. If an ammo bearer fails to bring ammo or a loader fails to load, or a gunner fails to fire, a weapon is lost and so may a mission be lost. Heroes are often made (if noticed) when they overcome a missing link in the team effort during combat. Unfortunately, many of these heroic efforts occurred every day in combat but were never recorded, either because of the confusion of battle or the lack of survivors. This paragraph relates back to the theory expounded during basic training, whereby a soldier lost all individuality and could be molded into a single team concept.

CHAPTER 6

Shipping Out

I AM NOW ready to embark on my third train ride. Destination unknown, very "hush hush," the train windows were all covered. The whole battalion was restricted to the camp for four weeks. The unit insignia was removed from all equipment. Our guns were covered with cosmoline, a heavy grease, and then loaded on flat cars covered with canvas tarps. The unit boarded the train around 12 o'clock at night and we were off to Patrick Henry, Virginia, which was near Fort Eustis, Virginia. We never knew where we were going until we arrived.

My unit arrived at Patrick Henry at about 4 o'clock in the afternoon where we were met and put under guard by many Military Police. Everywhere we went on the post we were escorted (guarded) by Military Police. If any of us went to the latrine the Military Police went with us. When we went to the mess hall Military Police stood over us. When we were at Patrick Henry five days where we were given physicals, immunizations, new equipment, and issued the new Garand semiautomatic rifle call the M-one. This was a newly invented and manufactured 30 caliber rifle adopted by the Army to replace the slow firing Springfield rifle used during the first world war.

We assembled and disassembled this rifle many, many times. We even learned to disassemble and assemble it blindfolded, but we never fired it. We daily ran an obstacle course under the watchful eyes of the MPs, but they never participated in the running. One man in our unit broke his leg which kept him from going overseas with us. It was rumored he had broken it on purpose, so he wouldn't have to ship overseas. If I had known what lay ahead I might have been tempted to break my leg, but as a Corporal I now had to set an example which I tried to do, and I was not lucky enough to have a real accident.

On the fifth day we got up at 3 o'clock in the morning and under MP guard loaded onto the trucks, each man carrying a pack and two duffel bags, one bag stenciled with a "B" the other stenciled with an "A" (more about these bags later). We also carried our new rifle, bayonet, and gas mask. We were so overloaded we could hardly move.

We had received the new model steel helmet with plastic liners. This helmet was the most torturous piece of equipment ever invented. It was heavy, it made your head hurt, and it wore your hair off. It had a chin strap to be used only in parades. If you were near an explosion using the chin strap it would break your neck. The helmet had some good uses which caused it to be called a "steel pot." When we got overseas, soldiers washed in it, cooked in it, dug with it, and shit in it, thus the name steel pot. We avoided wearing it unless we were ordered to, or in the event we were being shot at.

We off loaded the trucks on a dock next to a Liberty Ship, which was a round bottomed, thin skinned the ship called a "sea sick machine." These were newly built cargo ships rushed into production to alleviate the shortage of ships in our Navy at that time. These ships were mass-produced with little thought given to their stability or seaworthiness. They contained no frills or comforts for their crews or passengers, and for my first time aboard an ocean going ship I was very unimpressed. I was even glad I had been rejected by the Navy.

As I struggled with two barracks bags, a full pack, carrying my new rifle and wearing my damned helmet, a Red Cross lady handed me a doughnut, a carton of milk and a five-dollar bill. I still don't know what the money was for. I was told it could be used to buy necessities like razor blades, toothpaste, and soap; however, these things were provided by the Army, so we had to find another use for the money, which we did.

Struggling up the gangplank a sailor handed me a slip of paper that told me my hold number, my bunk number, and the time I could be in the bunk. I got to my assigned bunk which was a canvas spread over a steel frame 2 feet wide 6 feet long and 18 inches below the one just above it. The one just above it was of the exact same dimensions. I was supposed to share this bunk with two other guys, that is why I was given the paper telling me what time I could use the bunk. We also had to stow six barracks bags and our other equipment in or under it.

This ship was loaded with two battalions, approximately two thousand men. I knew at once this was not going to be a pleasant cruise. The first couple of days we stood in a continuous chow line. After breakfast we lined up for lunch and after lunch we got in line for dinner. When you had any spare time, you could use your bunk if it was your turn. If it was not your turn, you had to stand or sit wherever you could find enough space.

After the third day out most everyone was sea sick. The chow line was not as long, and no one had to worry about their time in the bunk. Anyone who was too sick to get on deck lay in the bunk. The stench of vomit hit you as soon as you started down into the hold. So many of the sailors were so sick, the Army had to man their guns. Luckily, we never had to fire the ship's guns as we had never trained on the Navy's big weapons. However, I could have fired their machine guns as they were almost the same as mine. I never got seasick because I refused to go down in the hold.

I did find a use for the five dollar bill the Red Cross lady gave me. The ship turned into a floating crap game and that became my main source of recreation. We were told after three days at sea that we were on our way to invade North Africa, so money did not mean much to most of the guys. They could not spend it or mail it home, so they just gambled it away. I won almost $1,000 shooting craps which I really tried to hang on to since that was the most money I had ever seen at one time in my life.

About five days out to sea the ship put out a raft on a long tow line, on the raft was a frame with a sheet on it. A bull's eye target was painted on the sheet. All the soldiers that could get out of their bunks got to fire their new M-One rifles at this target. This was to be a familiarization exercise.

I was given a single 30 caliber round and an eight-round clip. I lay down on the fantail of the ship loaded the single round and squeezed it off at the target, the casing was automatically ejected. I then loaded the full clip and squeezed them off. About five men were firing at a time, and to this day I don't know if anyone ever hit that sheet, but we now knew how to load our rifles. Keep in mind we were on our way to invade

North Africa, not knowing what enemy we were going to encounter. But then we were not infantry and we were only going to shoot down airplanes with our antiaircraft guns and machine guns. I wondered sometimes if we could not hit a slow-moving bedsheet whether we could have hit a fast flying airplane. I found out later we could and some of the units I was in were very successful at downing enemy planes.

For almost all the trip I volunteered to act as a lookout and was given a pair of binoculars so that I could watch for enemy submarines. I never saw any submarines from my ship, but I was told that one of the other ships in the convoy was sunk. When I heard this, I was glad I had not enlisted in the Merchant Marines. I carried my blanket right up to the bow of the ship where I stayed the entire time except to eat or shoot craps. There was always a good breeze and no sickening odors up there, so I stayed there night and day even when it rained. Being wet was better than being sick.

When we arrived in North Africa, the real invasion had already taken place at Casablanca by our infantry and our ships sailed into the port of Oran which had been secured by the free French a few weeks earlier. We had it much better than the poor infantry that stormed ashore at Casablanca. At least we didn't get our feet wet.

Everyone should give credit to the infantry as they always had the toughest assignments. The infantry has always been the most needed and the least noticed of all the services. They never received the publicity the Air Force received nor the good food and quarters of the Navy.

Even though my unit did not storm ashore with guns blazing, we did unload that damned ship with much apprehension and perhaps a little feeling of relief. Being penned up on that stinking ship for almost three weeks certainly was no pleasure cruise and some of the men were still so sick they had to be carried off on litters. I never got seasick and I had accumulated several dollars, so in reality I had taken an adverse situation and had at least turned a profit. I believe out of everything bad some good can be found.

CHAPTER 7

Oran North Africa

MY UNIT debarked in Oran North Africa at about 10 o'clock in the morning and at about three that afternoon the Germans sank the ship in the harbor with all our guns and heavy equipment still aboard. I don't think any of this equipment was ever recovered. Even if it had been recovered it would have been useless after the bombing and being submerged in the salt water.

My unit bivouacked on the hill outside Oran which we named Goat Hill. It was almost solid limestone. We set up pup tents and I dug my first combat foxhole in solid limestone. It took me three days and 14 blisters to have a hole 2-feet deep and 6 feet long, just big enough to lie in during an air raid which we received almost daily. All we could do was jump in our holes as we had no weapons with which to shoot at the planes. We did have our M-One rifles which we sometimes used without results; remember we cannot hit a bed sheet.

We camped on Goat Hill for two or three months awaiting new equipment. Most of the time was spent walking, guarding, digging latrines called "slit trenches" by the officers who had special designated ones, and was called "shit trenches" by enlisted men. There was a great need for these trenches because everyone suffered from diarrhea.

We walked double guard all around our area at night. One man looking forward and the other watching to the rear, for we were told the Arabs would slip up on a lone sentinel and knife him. We were told the Germans paid the Arabs for the ears cut off allied soldiers. This may have been a lie but it was effective in keeping us alert. Later I heard the British adopted the ear method for their Senegalese troops, who were black savages with a reputation for being able to slip into a two-man foxhole, taking the ears of one man without disturbing the second man. This put terror in the hearts of Italian and German troops. I think this practice was discontinued because it was impossible to determine whose ears were being taken, since German, Italian, British, or American ears all look the same.

We walked guard with loaded weapons and if we saw anything moving, we called, "Halt, give the password" and if they failed to give the password we were supposed to shoot. Every night you heard shots being fired and I was told some of our guys had been shot by our own people.

I remember one night I was walking guard and I saw a man moving in my direction when I called, "Halt, give the password." The word came back, "I'm the Captain and I got to shit." Right there he pulled down his pants and did his business. He had diarrhea so bad I don't guess he could remember the password nor cared if I shot him.

All the rations we had in North Africa were "C" rations or "K" rations. "C" rations consisted of two little tin cans. One can contained hard crackers, a small tin of powdered coffee, sugar, and a small package containing five cigarettes. The other can contained either beans, hash, or beef stew. This was the worst crap that could be found. The beans were briny with salt, the hash smelled like it was rotten, and the stew was all fat. I saw a starving dog refuse to eat the "C" rations. I think some of these rations were left over from World War I. Today at the very mention of "C" rations I almost vomit. Even though in the '50s the service supposedly made vast improvements in them, I still could never eat this stuff even if they have been improved.

We ate the "C" rations hot, cold, or not at all. I traded mine to the natives for eggs, vegetables, melons, grapes, or anything else I could eat raw or cooked in my helmet. I even tried Arab donuts cooked in goat grease which weren't too good, but better than "C" rations. The most valuable item a soldier could get was a one pound can of black pepper. If you put enough pepper on the "C" rations you could get them down, which would at least give your diarrhea something to act on. Too much pepper could leave you with a burning rear end.

"K" rations came in a little brown box covered with paraffin so that when the paraffin was lit, it would heat the contents. "K" rations were labeled breakfast, lunch, and supper. Breakfast had a tin of an egg concoction, hard biscuits, jelly, coffee and pieces of hard candy usually lemon drops. Lunch had a tin of cheese, coffee, hard biscuits, and a hard chocolate bar, which became a valuable trading item. Supper contained a ham spread, hard biscuits, coffee, and cigarettes, also good trading material. I never received very many "K" rations as they were issued to infantrymen because they were lighter and easier to carry. Maybe the cheese helped the infantry control their diarrhea. When we were issued "K" rations one person might get all breakfasts, another all dinners, or suppers. You never got all three different meals at one time. I talked to infantrymen who hated the "K" rations as much as I hated the "C" rations. None of the rations ever varied—the same meals day after day.

All the water we had to drink contained Halazone tablets, which (if you could get them) we would put the heart lemon drop candy in for flavoring in order to drink it. The Halazone tablets were purifying agents which gave the water a terrible taste that even with the lemon drops for flavoring was barely drinkable. Most everyone drank coffee to get liquid into their system, I have even used coffee to brush my teeth. The coffee was heated in an aluminum canteen cup which would burn your lips even after the coffee was cold. Many soldiers traded with the Arabs for wine, which the Arabs always referred to as champagne. Some of the wine must have been aged at least a week. I am sure this wine aided the diarrhea, but it was a liquid and we needed liquids to keep our bodies from dehydrating.

I have already mentioned some of the trading that took place, which in the beginning was not too rampant but later grew into the huge black-market operations that flourished in the European Theater of Operations.

I mentioned earlier that each soldier had two barracks bags. These were either blue or brown denim and had a rope drawstring at the top to close them. They were stenciled with the soldier's name, serial number, and either a big "A" or "B." The "A" bag was used carry to your combat gear to a forward area and the "B" bag was used to store personal articles in the rear area. The "B" bag became a valuable trade item.

The Arab men wore baggy trousers because they believed their Savior would be born to man and the baggy trousers were used to catch him (so I was told). The Arabs liked the barracks bags to make baggy trousers. They would cut two holes in the bottom for their legs and use the draw string as a belt. It was not unusual to see in an Arab running around in a barracks bag with a big "B" stenciled on his rear end, hence the saying "Go blow it out your 'B' bag," this many GIs will recall.

Another item of issue was two cot-sized mattress covers, one you could use as a ground cover or fill it with straw when it was available and use it for a mattress. The other was to be used as a body bag in the event of your demise. Many soldiers would trade one of them to the Arabs, who cut a hole in the top to put their head through and cut the corners off for arm holes, thus making a long robe. Even though it was extremely hot during the day in North Africa, the Arabs wore heavy long robes to retain their perspiration creating a dampness which cooled their body. I guess this procedure was comforting, but 'oh' how bad they smelled. I also heard that if a soldier was killed and he did not have a mattress cover, $1.80 was taken from his final pay for its replacement. I never saw a real sleeping bag or body bag until I was recalled for duty in the Korean Police Action.

CHAPTER 8

Combat

THE UNIT in which I was assigned, finally received new equipment and was converted to an antitank unit. We were given half-track vehicles with antiaircraft guns mounted on them. We were sent into action to help stop the German break through at Kassarine Pass. The antiaircraft weapons were totally ineffective against the German Tiger tanks and my unit was nearly wiped out during the retreat (which was referred to as a strategic withdrawal).

The members of my half-track crew and I were ordered to fall back to a regrouping area. This was the last order I ever received from my headquarters as all communications were lost. When I arrived at the regrouping area we were told by Military Police to keep moving to the rear and not to stop. In all the confusion it seemed no one wanted us nor could tell us what to do or where to go. The crew and I drove the track toward Algiers. We were out of ammunition and nearly out of gas when we arrived outside Algiers. We abandon the track when it ran out of gas and walked into Algiers. After four or five days in Algiers I was apprehended by the Military Police as I was coming out of the Casbah, which was off-limits to GIs. They took my belt and shoelaces and put me in a civilian dungeon which the military had taken over. I guess the reason they took my belt and shoelaces was to keep me from hanging myself which I would have never done, even though I thought about it a few times.

There were many other soldiers in the same predicament I found myself. I was told by the MPs that I was going to be tried for desertion in the face of the enemy and would probably be shot. Needless to say I was scared half to death. After three days in that dungeon I was taken before a Major who was conducting what was called a deck court-martial, which was set up to handle the many soldiers who had become separated from their units during the retreat at Kassarine.

The Major was very polite when he told me my unit could not be located, so he was only charging me with one day of being absent without leave. I was reduced to the grade of Private, fined $45 a month for six months, reassigned to another antiaircraft unit, and restricted to a gun site area on the roof of a warehouse on the docks at Algiers. We had a 40-millimeter gun and crew, plus a 50-caliber machine gun which one other Private and I manned on top of the warehouse. We had many opportunities to shoot down enemy aircraft as they flew over us almost daily. This unit was credited with shooting down and destroying at least twenty aircraft.

The warehouse contained Lend Lease material which when we cut holes in the wire screen barrier was found to consist of gallon cans of butter, powdered milk, powdered eggs, cheese, peaches, bacon, chocolate bars, and good pork and beans. Of course, we helped ourselves to these goodies until we got caught.

Rather than court-martial the whole crew, the Battalion Commander made us each donate $10 to the Red Cross. He said this would punish us without creating an international incident since the items we ate were Lend Lease intended for our allies. It is still a mystery to me how shipping space for these Lend Lease goodies could be found when all the troops could get to eat was "C" or "K" rations. This is just one example of how a political operation can take precedence over a military operation. As a civilian I could have questioned this program. I might not have received an answer, but I could have voiced an opinion. As a soldier, now a Private, I was not permitted to inquire about high-level strategy. All I could do was pay and obey.

CHAPTER 9

Cape Bon German Evacuation

THE KASSERINE Pass breakthrough had been repelled and the Germans were evacuating Africa at Cape Bon near Tunis. The antiaircraft unit I was with was dispatched to secure the area on Cape Bon. The enemy had abandoned all their equipment on the beaches, and almost every American GI had a motorcycle, a Volkswagen, (which was the German version of the American Jeep) or some other vehicle which they would ride around on while hunting souvenirs. We had more accidents and injuries from this abandoned equipment than we ever had in actual combat. It must have looked funny seeing a soldier driving a big German eight-wheel truck over the beaches looking for souvenirs. We would drive these vehicles until they ran out of fuel then abandoned them and find another one that had fuel and would still run.

I personally found a case of Luger pistols. I would strap one of the pistols on and go to an air base where I would sell it for $100 to an airman, always giving the same story that I had taken it from a German Colonel that I had captured. Up until that time I had never seen a German Colonel much less captured one. But the airmen believed anything they were told. Those same pistols that I thought I was getting big money for at that time, would bring $1,000 each on today's collectors market.

Cape Bon was a desolate stretch of sand and my unit was constantly firing at enemy aircraft coming from Sicily (which you could see with binoculars on a clear day). The German target was the newly built U.S. airstrips near Bizertie. A GI song was created about Bizertie, I don't remember the words to it but the title was *"Dirty Gertie from Bizertie."* This may have had something to do with the dirty Arab prostitutes who gathered around the air fields. All the ones I saw were ugly, dirty, smelled bad, and had black teeth from chewing beetle nuts. Most of these prostitutes were cast-off wives of the Arab men. Under the Arab customs a man could have as many wives as he could support. Arab women, like women in other societies as well, were considered inferior people. In Africa, a wife or wives had to walk behind their husband to show that they were not his equal. I have many times seen an Arab man riding a donkey while his women followed on foot carrying heavy loads on their heads or backs. In order to obtain a divorce, all the man had to do was tell a wife to leave his home. When this happened, which was often, the women usually turned to prostitution to survive. They would prostitute for anything they could get, a candy bar, cigarette, or even "C" rations.

On Cape Bon there wasn't much for us to do except man our antiaircraft guns and clean them when we were not firing at enemy planes. We manned the guns and cleaned them with a skeleton crew except during an air raid. The rest of our time was spent hunting souvenirs or trading with Nomadic Arabs, unless an officer came by and ordered us to fill sandbags to put around our guns which we did until he left.

One time I bought five fresh eggs from one of the Arabs for $40 which was a lot of money to pay for eggs. I never would have spent that much money but it was German money that I had found. The dumb Arabs didn't know it was no good after the Germans had left. I even traded some "C" rations for two chickens which I cooked in my helmet. It would have been a very boring time if it were not for the air raids.

After a month on Cape Bon the unit I was in started trucking us, a few at a time, into Tunis for a weekend pass. Most of the men would drink and hunt women or do both. I will not tell what I did, suffice it to say I went sightseeing and took pictures. A pass to Tunis gave us the chance to see how the French and Arabs lived in that part of the world. I met a man who was in the French Foreign Legion and I could not get away from him. I made the mistake of buying him a drink and he followed me everywhere mooching drinks. After I bought the third one for him, I finally managed to slip away from him. He reminded me of Little Willie who I wrote about during my basic training days. I had found a German Leica camera which still had film in it so I took pictures. The pictures did not turn out, as the camera had been exposed to the desert heat too long. I later sold this camera to an airman at a bargain for $50.

When we started to leave Tunis the Lieutenant who was in charge of the pass detail placed a case of wine in the back of the truck and told me I was in charge of it. Officers rode in the front of the truck and could not see what went on in the rear. As soon as the truck had started, I really took charge of the wine. We drank the whole case which I thought was the appropriate thing to do under the circumstances. A week before we went on pass the Battery Commander had issued an order that no alcoholic beverages were allowed in the battery area. I was helping to comply with that order.

I don't remember arriving at our gun site I was so drunk. Someone drug me off the truck and left me lay all night in the first rainstorm we had seen. This was the first and was to be my last big drunk in my lifetime. The next day I was wishing I could die which should have been punishment enough, but the Captain had other ideas about punishment. He made all the soldiers and myself dig a hole big enough to bury the truck. We did this after 6 o'clock each day until it was finished. The Captain, being a practical man, used the hole to bury our mess garbage and then we had to throw the dirt back on the garbage to cover it.

Since I had been in charge of the wine I was accused of instigating this episode and I was given additional punishment. I was put on KP duty for an indefinite period of time. The only time I could leave the mess area was during an air raid to man my machine gun.

The battery mess was in the open and consisted of gasoline fired camp stoves and huge square or round pots and pans. I had to fill the stoves, pump them up, light them and clean them after each meal. The pans were about three feet square made of heavy aluminum. Each pan would hold five or six cases of "C" rations. I had to open all the little cans of beans, stew and hash, and dump them all together in a pan to heat. This made a slop fit only for hogs to eat, but that was all we had most of the time. The only thing we had for bread was the hard crackers that came with the "C" rations. Occasionally, we were issued powdered eggs, powdered milk, and cans of cheese which we mixed with water and served for breakfast. The pots held 10 gallons of water which I filled to make coffee. This was about the only thing we had to drink other than the water that tasted like iodine.

Before each meal I had to dig a trench a foot deep over which we would sit three 25-gallon galvanized garbage cans. I filled all three with water and put GI soap in the first one. The other two were left with clear water. Next, I filled the trench with gasoline and set it afire to heat the water. This was a dangerous task as the gasoline burst into flames with a big blast. I learned to stand way back when I threw a lighted match in the trench.

When the troops had finished eating, they would empty their mess kits then dip them in the soapy water and scrub them with a brush. Next, they dipped them in the first clear water can which soon became soapy and then in the second clear water can. Today, we would worry about getting lead poisoning from the galvanized cans but at that time the only lead poisoning we worried about came from a bullet. Even after dipping the mess kits in the third can, I'm sure some soap was left on them, but this only helped hide the taste of the food the next time we used them.

I did this KP duty so long that the mess Sergeant wanted to make me a cook, but the Captain said I was needed on the machine gun. He finally relented and released me from KP. I don't know if he thought I had learned a lesson about partaking of officers' wine or whether he thought the KP duty was too easy. But whatever the reason, I was glad to be relieved of that dirty duty.

CHAPTER 10

Replacement Depot

AFTER THE invasion of Sicily, my unit was returned to Algiers for reassignment, but I had been bitten by an Arab's dog and had become very sick. I was evacuated as a combat casualty, given morphine, tagged through an aid station, sent to a clearing station, given more morphine and then evacuated to a field hospital where I was given even more morphine and most of the time I was in a semi-conscious state. When I got to the field hospital I was diagnosed with acute appendicitis nearly ready to burst, complicated further by yellow jaundice. I was operated almost immediately. In WWII, an ampule of morphine was in a bandage package carried by each soldier which he would administer to himself if wounded. Therefore, at each stage of his evacuation he was given much more morphine. So many men became addicted to morphine that it was removed from the first aid packets carried by each man. This was to prevent the morphine from being stolen by the men who had become addicted to the drug.

After I became conscious following the operation, the doctor asked me about the wound on my leg. When I told him I had been bitten by a dog he began giving me the Salk vaccine shots to prevent hydrophobia. These shots were very painful. They were given on0e a day in the stomach where I was already sore from the appendectomy. I was to get a shot for 38 days. After 10 days, the doctor took pity on me, loaded me in an ambulance and accompanied me back to Cape Bon where we found that mangy cur dog still alive and mean as hell. After the dog had checked negative for rabies, the doctor could then stop the Salk vaccine which seemed to relieve him and it damned sure relieved me.

While I was in the hospital Bob Hope made his first visit to a combat zone. I was not an ambulatory patient; therefore, I did not get to see the main show. However, he and the other performers came into the hospital tent where I was and put on a little special performance. Francis Langford was there with Bob Hope and she sat on my cot and sang, *"You Belong to Me."* It's still my favorite song and I am one of the few guys that can brag about having been in bed with Francis Langford, a most beautiful lady who did much for the morale of the soldiers. If she ever reads this book, I would like her to know that mere words cannot express my appreciation for the special attention.

When I was ready to leave the hospital, I was sent to a replacement depot. The replacement depot was a hobo camp set up in the desert near Bizerte. It consisted of two-man pup tents set up for shelter from the sun, a long mess tent with a big bulletin board where orders were posted daily, and a supply tent where you could trade your uniform for one that had been laundered. Many of the exchanged uniforms had bullet holes in them that had been repaired after they had been salvaged from the hospitals. If you were caught for reassignment, you would be issued a minimum amount of used combat equipment before you were sent to your next unit.

There were so many soldiers there that it was impossible to keep track of them. I thought Camp Stewart was bad but compared to the replacement depot it was heaven. The big task of all soldiers there was to keep from being noticed by the people in charge of making up shipment rosters.

Most of these men had been wounded in combat and evacuated through a hospital. In North Africa a soldier could only be shipped home if he had lost an eye, an arm, or a leg. Most of the men who had seen combat where not in any hurry to see more. Consequently, hide out camps were created on the outer fringes of the main camp where men could escape being noticed.

One of these hideouts was in a deep gully where the guys were living like hobos, trading with the Arabs for food which they cooked in helmets and wine to drink. One night while I was there, about 20 GIs in one of the hideouts acquired some poison wine which they drank. When it was reported to the Depot Commander, they were all shipped to a hospital. I never saw any of them again, so I never learned whether they lived or died. I didn't drink any of the wine because of my wine drinking experience on Cape Bon.

Well, I finally got caught and was put on a shipping roster. I was assigned to an antiaircraft unit located in Algiers that was on alert for the invasion of Southern France. They were going to a staging area on Corsica, where many allied forces were being assembled.

While waiting for ship transportation for the move to Corsica we were given passes in Algiers. The Casbah (native quarters) had been placed off limits and we were not allowed to enter these quarters. It was rumored that some GIs had been murdered in the Casbah by the Arabs. The Arabs then cut the testicles from the GIs and had sewn them in the dead men's mouths.

This was reportedly done because the soldiers had molested the Arabs' women. I think this rumor was started to cut down on the many cases of venereal diseases that were being transmitted to the soldiers. It seemed a high percentage of the Arab prostitutes were infected with many types of venereal diseases, most of which I had only heard about by watching the GI movies shown to us during our training in the States.

While General Patton was in command of Algiers, GI houses of prostitution were established to aid in stopping the spread of venereal diseases. If a soldier wanted to get laid, he could go to one of these authorized houses were certain rules had been established for their protection.

All the women were inspected on a daily basis, and when a GI entered the house he would pay two dollars to a guard on duty at the door. The guard would stamp the man's hand to show he had paid. The soldier would then be given a "short arm" inspection by a medic and allowed to proceed if he showed no sign of VD. Next, the GI could take his pick of the prostitutes and go to a room with her where they did their business. When he had finished he went to a latrine where he washed up and was handed a prophylactic kit which he was forced to use.

This Pro kit contained sulfa pills, two tubes of ointment, issue paper and a small cotton bag with a drawstring. The procedure for using the Pro kit went like this: take the sulfa pill, squeeze one tube of the ointment into the penis, rub the other tube on the outside of the penis, wrap the tissue paper around it and put the cloth bag over the paper to hold it in place. This procedure might not prevent you getting the disease, but it did keep the ointments from staining your uniform.

It was rumored that General Patton said a man that wouldn't screw wouldn't fight, and he wanted his men to fight. It was also known that a man with a venereal disease was unable to fight with full effectiveness. Venereal disease has always been a huge problem for the military services.

My reason for telling about the GI houses of prostitution is to acquaint you with an incident that I caused in Algiers. What was meant to be a joke turned into a small tragedy which I will always regret.

In this new outfit I had joined, I was assigned an assistant machine gunner. He was a new replacement fresh from the States. His home had been in Alabama, so I'll give him the name of Al. He was a young, naïve kid that had just graduated from high school. He claimed to be a virgin and all the guys kidded him about getting laid. When he went on pass and came back he was bragging about this Arab girl he had met. He told everyone he had laid her and when I asked him if he had used a Pro kit, he didn't know what I was talking about. I gave him one and told him to use it, but I didn't tell him how it was to be used. The directions on the kit said to leave it on for two days, which he did. He didn't even take the bag off to urinate, but urinated through it. The urine caused the draw string on the bag to shrink which cut off the blood flow to his penis, and his pecker started to swell. When he told me what he had done, I told him he had waited too long to use the kit and had probably caught the bull head clap. I sent him to the medics where they took the bag off and found he did not have a venereal disease. I was reprimanded for this joke which did not bother me much until later when I heard Al had been killed in an air raid on Corsica.

50 caliber machine gun on top of warehouse in Algiers North Africa – 1942

CHAPTER 11

Airborne and Sicily

DURING THIS period of time the Army issued a call for volunteers to join the Airborne. Parachute jump pay was an extra $50 a month and they wore some new fancy uniforms. I had been demoted to Private and I could use that $50, so I volunteered. I had been told that a person should never volunteer for anything in the Army, but I had not yet learned that lesson. After volunteering for Airborne, I never volunteered for anything again.

I was accepted for the Airborne and released from the antiaircraft unit that was going to Corsica. I had not made many friends in that unit anyway and they were glad to get rid of me.

I started jump training by putting on a parachute with the help of an already airborne qualified Corporal. This part wasn't too difficult. We next jumped off the tailgate of a 2 ½ ton truck while it was still standing. We would yell "Airborne," jump and roll on the ground. I never learned what the yelling had to do with anything, but it was required of us. When we had mastered the technique of jumping from the truck standing still, we advanced to the next phase. We now jumped from the truck while it was moving at a speed of 20 miles an hour, and even at this speed we still had to yell, "Airborne."

I heard a tale later that the yelling was meant to build your confidence. Some units supposedly would yell, "Geronimo," after the famous Indian warrior of our old Wild West. I also heard that "Geronimo" evolved from troopers yelling, "I don't want to go" but I don't know if any of this is true.

While participating in this ground training phase, I learned there was no danger in jumping but you could sure as hell get hurt on the landing. After two weeks of jumping from trucks, we put our chutes on and went up in an airplane and JUMPED out. When we landed we were qualified paratroopers.

Less than a month after I became a paratrooper, I joined a unit that was to make a jump in Sicily. We were told that our Army was stalled along the coast of Sicily and we were going to jump into the German rear area. We would then blow up German supply lines and disrupt their lines of communication.

That was the grand plan, but this is what happened. We loaded aboard the plane alright, but from that point on everything went wrong.

The plane I was in had some engine trouble and fell behind the other planes in the formation. When the airplane got near the coast of Sicily, I think the pilot panicked. In the back of an airplane used for jumping are two lights over the jump door, a red light which means don't jump and a green light which means jump. Troopers are taught to jump automatically when the green light goes on.

When the pilot panicked, he hit the green light and we all went out. The plane was about three miles short of the coast and as I was one of the first to jump, I landed in the ocean.

Some of the last troopers to jump actually made it to shore. Four of us that landed in the ocean, thanks to our lifejackets, were able to reach a coral reef where we were stranded for four days with only the rations we carried and our canteens of water.

We saw Sicily, but only from a distance of about three miles. On the morning of the fourth day, a U.S. Navy boat picked us up and we were returned to a hospital in North Africa. We were treated for cuts, bruises, and exposure. This was to be the end of my airborne service until much later in my career.

The jump by the airborne unit must have accomplished some good as the American forces advanced rapidly up the coast of Sicily. The Sicilian Campaign soon came to an end and our forces moved into Italy.

I read later, after the war, that one reason the Americans had achieved so much success in WWII was due to the fact that our forces were so confused about tactics that the Germans could not figure us out. There are rules of tactics that the Germans followed with strict obedience. Our civilian type Army had never learned of these rules and were always doing the unexpected.

Individual actions many times altered plans and turned a failing situation into a successful operation. Many of these individual actions, if observed or recorded, resulted in the participants being rewarded with medals for valor. Unfortunately, many, many of these actions were never noted or rewarded, either because there were no survivors or because of all the confusion during the operation. It has been said that some of the actions that were rewarded could have just as easily resulted in disciplinary action if they had turned out unsatisfactorily.

CHAPTER 12

Naples, Italy

BY NOW, Allied Forces were in Italy and when I was released from the hospital I was sent to Naples to another replacement center. This center was better organized than the one to which I was assigned in Africa. It was housed in a college Mussolini had built that had not as yet been bombed.

We were issued blankets and assigned to rooms with 50 men to the room where we slept on the hard marble floor. The people in charge could keep better control of us since there was little room to hide and we had to check in and out with a Sergeant each time we left the room.

We were given day passes so we could look around the town and visit a Red Cross Service Club. This club was across the street from the Naples Post Office which was completely destroyed by German bombs while I was there. I think the target was the Red Cross Service Club, but maybe even the Germans did not always hit the right target. Anyway, the service club was untouched and was a place where we could read books, play games, and have coffee and doughnuts.

While I was in Naples the fighting was taking place at Monte Cassino and replacements were being sent by the hundreds. I was really sweating it out. However, I never made the levy because of my experiences in Africa, the Sicilian fiasco, and having been hospitalized. At this time I only weighed 110 pounds and even though there was an airborne unit in the area I doubt if I weighed enough to pull a parachute down. The extra $50 a month no longer interested me.

I contracted malaria and was hospitalized for more than a month. This hospital was in a building that was clean, but hot. We had steel cots, mattresses, sheets, and mosquito nets. This was the nearest thing to a real bed that I had been in for over a year. Even in the North African hospital we only had canvas cots. I was given huge doses of atabrine, a synthetic quinine which helped to put my malaria into remission.

I heard some mention of being sent home, but since I could still walk it never happened. I was sent back to the replacement depot where I was marked for limited duty. I was given regular passes which gave me the opportunity to see Naples and some of the horrors of war suffered by the civilians trapped in a war zone.

There were little kids by the hundreds with no families, many of them lame or scarred from injuries. They had no place to go, nothing to eat, ragged clothes and they slept wherever they found room, mostly in the streets, doorways, sewers or the subways. They begged, prostituted, or shined shoes, anything to survive. I thought I had endured a rough childhood until I saw these kids. Never again did I feel sorry for myself. These kids made me realize that when things are bad they can get worse. The standard query from the young beggars was, "Hey Joe, what you want, a-nice-a-girl, you-take-a-my sister, she do anything you want." A soldier could have a girl all night for a chocolate bar or a package of cigarettes.

The black market was rampant, you could sell a package of cigarettes for $20. A pair of combat boots brought over $100 and the Italians would even by C rations to eat. On the black market you could get $100 worth of lira (Italian money) for five gallons of gasoline. A soldier could sell anything he had because the war had left the Italians with nothing.

In an attempt to cut down on black marketing the military started issuing script, "phony money," we called it. Servicemen were supposed to turn in all their U.S. currency and most of them did, but some sold U.S. currency for four times its value in lira.

I sent my entire pay on to my grandmother who deposited it in a bank for me. I lived on the money I received selling my cigarette ration which amounted to a carton a week plus the cigarettes from my rations. I don't smoke to this day, so maybe I got some benefit from the cigarettes besides just money. An enlisted man could not send the money home unless he had an officer certify that he had gotten it legally. I won almost $1,000 in a crap game in the streets of Naples and had to give an officer $10 per hundred to have him certify I came by it legally and send it home for me. I never did anything like this after I was commissioned.

I could have stayed in Naples forever. The climate was great, it was located on the beautiful scenic coast of the Mediterranean. The people were nice, the girls beautiful, and it was very historic. I became acquainted with several nice families, I gave them my rations and they shared their food with me. I think the Neapolitans were the friendliest of all the Italian people I met.

I met a beautiful girl guide who escorted me to the ruins of Pompeii and explained why the ancient city was destroyed.

Her version was that the gods on Mount Vesuvius became angered with all the immorality going on down there, so they exploded and sent an ocean of lava down to bury the city and its people. My guide showed me a part of a building that had been a restaurant and bar in the old days of Pompeii. On one side was a room called a vomitorium which had a sand floor. She said the old Italians went to the restaurant to eat and drink. When they had feasted on all they could hold they would go to the vomitorium. There they would stick a finger down their throats, empty their stomachs, and then go back and eat and drink some more. She said some would repeat this process over and over every day. I guess the vomitorium was like today's toilets.

I was next shown some engravings that looked like a penis with wings on it. My guide said these emblems pointed to houses of pleasure. We went into some of these ruins where I saw paintings on the walls depicting the specialties of the house. These pictures would be considered pornography in the United States, but in Pompeii it was ancient art. She said the paint called Pompeii Red had never been duplicated. If ever you have the chance, be sure to visit these ruins.

While in Naples I was getting stronger and I was able to find several places where I could get fried eggs, fried potatoes, bread, pizza pie and Vino. This food was about all that was available and it was expensive, but with my cigarettes I could afford it.

I spent a lot of time near the docks of Naples where I could get beer and peanuts and listen to Italian songs played on a violin or concertina. I was beginning to learn the language and getting acquainted with the people. The population consisted mostly of old men, ladies, girls and young boys under 15 years of age. Most of the grown men and boys had been killed or were in prisoner of war camps. Thousands of Italian prisoners had been captured in Africa. Contrary to what was to be expected the Allied Forces were welcomed in Italy by most of the people. Maybe the reason we were welcome was because we had so much, and they had so little.

CHAPTER 13

S.I.A.M. Company

J STAYED around Naples doing odd assignments until after the Anzio break-out and the bypassing of Rome. One of my assignments was guard duty at a gate to a villa on the outskirts of Naples where a top-secret unit was being formed. This was an easy assignment; all I had to do was check the identification cards and salute the officers.

While guarding the gate one day, a Major stopped and asked me if I would like to be assigned to his unit. Since almost all the people that entered the villa were officers, I wanted to know more about what I was getting myself into, so I asked him what kind of unit it was. Ever since volunteering for airborne, I was cautious about volunteering for something I knew nothing about. He told me all he could say was that it was a S.I.A.M. unit. When I asked what S.I.A.M. stood for he said, "Society of Inebriates and Misfits." I felt like I would fit into a unit of that nature. I had been inebriated on occasion and I was feeling like a misfit. Then he told me it was a Signal

Intelligence Company. I felt even better then, because I thought I was intelligent enough to stay out of the airborne, infantry, or artillery. At that time I wanted nothing more to do with a combat unit and I figured a signal intelligence unit would never be closer to the front lines than Army headquarters, which was a pretty safe place to be.

I told the Major I would be happy to be assigned to the company if he could manage it. About a week later I received my orders and instead of just guarding the gate, I was allowed for the first time to enter the villa. I was introduced to several of the officers whom I had passed through the gate and they all seemed to recognize me.

The new unit was being formed for a specific mission and it took some time to gather the skilled, mature, intelligent men needed. This unit had priority on all men coming through the replacement center. I probably would not have been accepted had the Major not noticed me guarding the gate, as I was neither mature nor intelligent.

This unit was being formed as a large separate company commanded by a Major under the direct control of Fifth Army Headquarters. The Major in command tried to get a mix of men who had seen combat action and men who could learn quickly the skills needed for the mission. I fit the combat action part. This unit had 32 officers, mostly lieutenants. This was the first time in the service that I was in direct contact with commissioned officers. Always before enlisted men had to use the chain of command to speak to an officer. I learned they were not from another planet but pretty much regular people. There was far less formality between officers and enlisted men in this company. From the first day I was there I sensed a feeling of dependability from the Major down to the lowest enlisted rank.

The Major had been a college professor and all the other officers were college graduates. Many of the skilled men taken for this unit were older and had been in business for themselves before being drafted. I remember one of them had been a banker, another a CPA, and one had even been a lawyer. The lawyer could have been commissioned but he did not want the responsibility. I had only graduated from high school and at first I held them in awe. Because I had survived in combat they seemed to rely on me more than I did on them and we established a mutual relationship. I was the youngest person in the company but at times I felt I was more mature than the oldest man. I had faced the unknowns of combat and many of these older men had just been drafted and sent overseas. They kept asking me what combat was like, but I could not adequately tell them. I don't think anyone could ever really describe life in combat. You can relate certain instances but to describe all the horrors of war is impossible.

By the time my new unit was fully staffed Monte Cassino had been obliterated, our army had broken out of Anzio and bypassed Rome. We could now move north up the peninsula without much danger from ground forces. We were still at the mercy of the German Air Force which had been severely curtailed in its usage in Italy at this time. We were strafed only once on the way to Florence. During the strafing run we were lucky that we had no casualties and the German planes only made one pass at us. I guess they thought a small convoy such as ours was not worthy of a second pass or maybe they were getting low on fuel. This action really only broke the monotony of the long ride. This strafing run was the first introduction to most of the men of what it was like to be shot at. Thank God no one was hurt.

CHAPTER 14

Florence, Italy

THE UNIT was loaded onto trucks and we were off to Florence. As we passed Monte Cassino I could see the total devastation created by mass bombings. It was said our Air Force had small fighter planes directing traffic for the bombers.

One man I talked to in our unit had been there. He told me that three times bomb concussions had lifted him out of his slit trench and he was not very happy with our Air Force. This man claimed he even fired at our planes with his rifle. I don't know if this was true but I can understand his feelings. Our Air Force had destroyed all the buildings, but before our infantry could climb the mountain the Germans were back in the rubble firing at them. This shows that the Air Force with their bombs cannot win a battle by themselves, but that it takes a combined force to overcome an obstacle.

The area around Anzio was covered with wrecked vehicles, guns, and other debris of war. Rome was declared an open city and all the Germans had withdrawn, thereby saving it from destruction. Allied combat troops were not permitted to enter Rome, so our unit went around the edge and I never did get to see Rome. When we arrived in Florence I saw dead horses and dead bodies where the armies had fought at the Arno River Crossing. The Germans used horse-drawn artillery and used horses and wagons to haul supplies. When we entered Florence there were few people left. Most of the civilians had fled to the countryside where they could hide on farms that did not present much of a target for bombs.

My unit occupied a large villa on top of a high hill. It had been a beautiful mansion and the lady that owned it remained there occupying a couple of rooms for herself and a couple of servants. I think she was a Duchess or some kind of royalty. During the two months I was there I probably saw her three times. She stayed in her room and a man servant took her food from our mess.

The villa was very large and counting the servants' quarters there were nearly a hundred rooms. Only one servant and his wife had remained, so we shared their quarters. This enabled each four men in the company to occupy a room. The man servant and his wife did most of our cooking which left us with time to train and study.

An intensive training program was immediately established by the officers who had been in training at the villa in Naples. They were now teaching the enlisted men how to operate high powered radios using both voice operation and Morse Code operation. We were taught to put messages in secret codes and also to decode the messages. We used the coding procedure on nearly all messages.

To operate with Morse Code took a certain degree of talent. One had to have very good hearing ability to distinguish between a dash and dot. I did not possess this capability, so I only had to learn voice operation. This left me with additional time off, which gave me the opportunity to explore Florence.

Florence, Italy is really the art capital of the world. It was here that the Angelico painting was on the ceiling of a cathedral. This painting is beautiful beyond description. There were other famous paintings and frescoes, which I saw in or on nearly all large buildings.

There is a covered bridge over the Arno River called the Ponte Vecchio, which is reputed to be the oldest bridge still standing intact in the world. This bridge was not destroyed during the war and it is considered a national art treasure by the Italian government.

Model statutes were everywhere. Many famous sculptors learned their trade in or around Florence and it is here that most of the famous Italian marble is quarried. Alabaster which is a soft form of marble also comes from Florence. Lois, my wife, and I later acquired some alabaster carvings which we covet.

During my time off, while roaming the streets to see the sights, I met a pretty girl with her brother near the Ponte Vecchio. I tried to speak to her in what little Italian language I had learned, and at first, she tried to ignore me. Had her little brother not spoken English I might never have become acquainted with her. He had been studying English and was anxious to try using it, which worked great for me. After her brother and I were conversing she started to speak to me. I learned then that she had also studied English but she seemed reserved and was rather hesitant to speak to me in English. When I tried to use my limited knowledge of Italian to talk to her the ice was broken and she became less hesitant to speak in English.

We talked for a little while, but then they had to leave for home as it was getting dark. There had been a 9 o'clock curfew established in Florence which meant everyone had to be off the streets at that time or were subjected to being shot. Before they left, they agreed to meet me at noon the next day at the same place on the bridge. They were going to teach me Italian and I was to help them with their English.

I was on the bridge a little before noon the next day. I was afraid they would not meet me as I knew nice Italian girls had been warned not to fraternize with American soldiers. I was very relieved to see them even though they were about 20 minutes late. We met and talked each day for the next three days, always with her brother present and finally they agreed to let me walk home with them.

There I was introduced to the rest of her family. Her father had been a city official of some importance and they lived in a nice villa which was heavily shuttered with iron doors and grills. Her mother was very friendly toward me even though she had lost a son in the war. Mearella (I will call her) had an older sister whose husband had been killed in North Africa. The sister seemed very lonely and she lived in another nice apartment several blocks away which I was asked to visit. I visited her several times and one of my older friends in the company became quite enamored with her.

Mearella was a very beautiful girl, 17 years of age, and she became the first girl I had a crush for. She seemed to like me a lot, maybe because I was the only male (other than American soldiers) near her age left in Florence. Her parents let me visit her in their home, but they would not allow me to be alone with her in public. We were always accompanied by her brother when we went for walks, so they became very good guides and showed me many of the sites I mentioned.

The friendship with Mearella and her parents grew to a point that we seemed like a family. I spent several nights with them to avoid being on the streets after curfew. They even allowed me to invite one or two of my buddies to their home. We would talk the mess Sergeant out of some excess rations which we took for her mother to prepare and we all shared many fine meals.

My grandmother sent me some corn to pop which we did at Mearella's home. This was a surprise to them as they had never seen popcorn before. They all enjoyed it so much I was obliged to have more sent to me, which I gave to them. This interlude to war in Florence and my relationship with Mearella was the most memorable of the entire time I spent overseas in World War II.

All too soon, this relationship came to an end as our training period was over and we were sent back into action. This was the end of my first love affair. I thought of Mearella many times but I never saw her again. I hope she survived the war and found happiness.

CHAPTER 15

Poe Valley

WHEN THE training period in Florence ended, we started our mission. We were sent to monitor the radio transmissions of the units along the front in the Apennines Mountains, which at that time was the most miserable place in the world. The Allied troops were stalled there in the rugged terrain for the entire winter. My company assigned a radio section to each of the front-line regiments and placed a section with the Infantry Division Headquarters. Our own company headquarters was located within the Fifth Army Headquarters area.

Our mission was to intercept all radio communications possible and relay these messages directly to Army Headquarters. We would also relay messages from Headquarters to the forward fighting elements. Many of these messages contained intelligence information and had to be coded or decoded.

I was with one of the front-line regimental sections alongside the infantry where we lived in foxholes. It rained, sleeted, or snowed every day. The foxholes filled with water and had to be bailed out daily. The roads were rivers of mud and when it snowed they became frozen ice rinks that our trucks could not move on. We were lucky to be in an area where a specially trained mountain division was operating. This division had mule pack trains which became our only means for receiving supplies. With all our many military vehicles we had to revert to the reliability of pack mules in the mountainous terrain.

I remember my mess kit filling with snow faster than I could empty it while eating. If we heated our chow it became cold so quickly that we stopped even trying to heat it. The only way we could keep warm was to put on all the clothes we had and exercise. We would dig a hole and fill it with gasoline then set it on fire. This was the only fuel that was available, and a gasoline fired did not last very long. This was a dangerous practice, but it was worth the risk just to be warm for a few minutes.

After we had been there a month, we were issued some gloves and rubble galoshes. When we had a chance to sleep we would try to stay dry under our ponchos which always leaked. We would crawl into as dry a hole as we could find wearing all the clothes we had. I would take off my gloves to remove my galoshes so my shoes would dry; then put the gloves back on to sleep in. If I had dried out when I awoke I would use my bayonet to scrape the dry mud off my trouser legs, put my shoes and galoshes back on, and start the everlasting attempt to keep warm.

During the entire period we were being shelled, strafed, or under attack by the Germans from the ground or the air. I don't know which was our worst enemy, the Germans or the elements. There is no way to adequately describe the misery of survival under which we existed. It's still a mystery how our soldiers survived that winter. George Washington could not have had it worse at Valley Forge. The only way anyone could understand these conditions is to have been there.

Just before spring we began to receive word through intelligence that random Italian partisans has started harassing the rear supply areas of the Germans. A decision was made at Army Headquarters that contact with these partisans should be established. Our unit having the best long-range communications system was given this assignment. Three-man teams were formed to infiltrate the German front lines where communications could be established between the partisans and our advancing infantry. This procedure was eventually to become the basis for the establishment of Special Forces (Green Berets).

I was assigned to one of the infiltration teams which was composed of one officer (usually a lieutenant) and two NCOs. The officer was in charge of the team and the NCOs operated the radios. We were told to stay in hiding and only use our weapons if we had to defend ourselves. This was not a "sneak and peek" spy operation. According to the Geneva Convention rules, if we were in uniform we would be prisoners of war and would not be shot as spies. We always wore our uniforms.

My team was infiltrated through the German lines hidden under a load of firewood in a donkey cart. This first trip went uneventful. I don't believe we were seen by the Germans because of the bad weather. We arrived in Bologna after crossing the Poe River where we were hidden in an attic by an Italian lady. While there we would radio any action that we observed to the front-line infantry units.

At the start of spring the push by the Allies began. The objective was to cross the Poe River and move through the Poe Valley. This campaign heralded the beginning of the end to the war in Italy. The crossing of the Poe started with tremendous concentrations of artillery and air attacks. Thousands of

German soldiers were killed, and the horse drawn artillery and supply wagons they were using were destroyed. The stench of dead bodies of men and horses was unbearable. I saw burned bodies hanging out of tanks and vehicles, still in the position they were in when their vehicles were destroyed.

I lay on a rooftop in Bologna where I watched the Germans withdraw. After the Germans left Bologna it was almost deserted. I radioed this message to our infantry, which speeded their advancement with little resistance. I could go in and out of the deserted buildings exploring the possessions the people had left behind. I never took any of these things as I had no way to carry them and I was afraid the things that were valuable might have been booby-trapped. The Lieutenant I was with was killed while aiding some partisans in protecting a bridge our infantry wanted to use in crossing a stream. This left me in charge to finish our mission since I was the senior NCO.

A few days after the Lieutenant was killed, I was bathing in a fountain in the courtyard of the villa in which I was hiding, when one of our airplanes strafed the villa. I was able to dive under a stone bench, which saved my life. The plane left bullet marks on the bench but left me untouched. That was the closest I came to being wounded in World War II by shell fire.

When our troops had secured Bologna, I was withdrawn to Army Headquarters, and there I was given a presidential appointment as Second Lieutenant. This made me an officer and gentleman by Act of Congress. A week later I was returned to the same job I had been doing as a Sergeant, but now instead of wearing stripes on my sleeve I wore a gold bar on my collar which I could remove in combat, thereby making myself less of a target.

CHAPTER 16

End of the War in Italy

WHEN I returned to the forward area north of Bologna, the German lines of resistance had become very fluid and in many places collapsed. The German troops were giving up by the thousands.

While driving down a road, I decided to test fire the machine gun mounted on my jeep. I fired a burst over what I thought was a vacant weed field. When I stopped firing the weed field became alive with Germans standing up from the weeds where they had been hiding. The Germans were holding their hands and their weapons above their heads wanting to surrender.

I was unable to accept their surrender and didn't know what to do. Finally, I told them to pile their weapons on the road and to form a column. We then found a white rag which we tied to a stick and gave to the leading man to carry. I told them to march on the road until they met some American soldiers who would accept their surrender.

We drove on ahead and the Germans started marching in the direction of our oncoming infantry. I don't know what finally happened to them, but I was no longer responsible for them and I assumed their days of fighting had ended. Shortly after this incident all the German Army that was in Italy surrendered.

Ever since the invasion of France at Normandy, the Italian campaign occupied a secondary role. It was even called the "forgotten theater of war" in some news releases, but I can assure you that any man that fought there will never forget it.

It must be remembered that the Allied Forces in Italy kept eight German Infantry Divisions from confronting the Normandy Landing. Our Allied forces also kept many German Panzer tank units occupied as well as many artillery battalions from firing on the invasion forces. All of this plus killing many German soldiers and siphoning off much critically needed supplies certainly was a major consideration in the success at Normandy.

After the Germans surrendered, my unit was returned to Largo de Garda where the unit was disbanded. Lake Garda is a beautiful area in the northern section of Italy not far from Venice, Switzerland and Yugoslavia. I had the time and was able to visit all these places. From Lake Garda we had a spectacular view of the Alps Mountains. Mussolini had built a beautiful home on an island in the lake to be used as a retreat area for him and his girlfriend. While the home had not been destroyed in the war, it had been looted by the Italian partisans.

A point system was established for the purpose of rotating the soldiers to the States. A soldier was given two points for each month he had been overseas and five points for each medal he received. Points were given for other things, but I have forgotten what they were. As it turned out, it made no difference in my case.

I had been overseas at that time for 36 months which gave me 72 basic points. For awards and other items, I had accumulated 111 points. This was the highest total that anyone had in my outfit but they all received shipping orders back to the States before I did. All my points did me no good as it was decided that the low point men should be shipped home first. They were to be given a 30 day leave and then shipped to the fighting in the Pacific which was still in progress. Men with the higher points were returning to the States to be discharged.

In the meantime, while I was awaiting shipping orders I was sent to the Italian Riviera to help run a combat rest camp. This camp was set up to entertain the high point combat troops while they were awaiting transportation to the States. The camp was established for enlisted combat soldiers only, and we had a sign erected at the entrance that read: "This Village Off Limits to Officers Unless Accompanied by Enlisted Men."

We had a platoon of Military Police assigned, whose only duties were to break up fights and help any man drunk or sober to his room. I don't recall there being many fights as these men had seen all the fighting they ever wanted during the war. I will not comment on the drunks.

The rest camp had taken over a complete village located on the beach of the beautiful Mediterranean Sea which is very near Southern France. The Army requisitioned and controlled all the hotels, bars, casinos, and cabanas on the beach. Each man at the camp was given a private room with maid service. The rooms had clean beds with sheets which were great luxuries after having lived in the mud, rain, or snow for more than a year. In my case, three years. The troops were given their meals free of charge in the restaurants of their choice. The food was provided by the Army and prepared by the owners of the restaurants. They were given free cigarettes, candy, and PX supplies. They could order anything they wanted to drink in the bars or on the beach at a low price.

This was truly a great rest area, but I don't recall seeing many soldiers resting. Everyone seemed to stay busy sunning, drinking, dancing, playing, swimming, or fraternizing with the ladies. There were many displaced persons migrating through this area. Many were women who were more than anxious to be entertained or to entertain our soldiers.

The Andrews Sisters had a big recording hit out at this time called, "Drinking Rum and Coca-Cola," which we played over loudspeakers on the beach. The troops really liked this song because we had plenty of rum and had received a ship load of Coca-Cola. A drink I liked was cherry brandy which we obtained in 5-gallon glass containers carried in wicker baskets. This was a great drink when mixed with Coke, but if you drank too much it left you with a big headache. This I know from experience.

Another song of the time which we played a lot was named, "Pistol Packing Mama." The ladies that passed through the camp would try to sing the song. With their limited grasp of English the song sounded more like, "Piss Hole Packin' Mommy," and the second line was "lay that piss hole down babe." Our troops got a real charge out of the ladies singing, and this song became the camp theme song.

About three months after the war ended, I was trucked back to Naples where I boarded another transport ship, not a Liberty ship. This time it was not as crowded as it had been on the Liberty ship I had gone to Africa on. I was not in charge of anything— no chow lines, no latrines to clean, no bunks to share, all I had to do was play bridge with the other officers. Officers didn't shoot craps and weren't supposed to gamble. However, we did have a few poker games in which I won a few dollars. I even shared a state room with three other officers and we had a latrine in the state room that was cleaned daily by the ship's personnel. This was more like what I thought a sea cruise should be.

It took almost two weeks to dock back where I left from three years earlier, Fort Patrick Henry. This time there were no Military Police guards but the Red Cross ladies were there. This time we were given doughnuts and fresh milk. It was the first fresh milk I had received in more than three years. The milk was much more impressive and welcome than the flying five-dollar bill we received earlier, even to a poor boy. The milk was SO much better than the wine, beer, cognac, brandy, coffee, and lemon drop halazone water I had been drinking, that I thought I could never stop drinking even if I burst.

I had a very brief stay at Patrick Henry, I was given a big steak dinner and spent the night in a clean bed which I didn't have to make. The next morning I boarded a train for Camp Atterbury, Indiana. It took a day and a night to reach Atterbury where I was to be processed for discharge.

Because I had suffered malaria and other sickness I was admitted to the hospital for observation overnight. An orderly would come into my room about every hour and shine a flashlight in my face, then ask me if I felt all right. After the third time he did this, I was mad, and I told him to keep the damned light out of my face so I could get some sleep. I guess he got the message because he never came back after that. I don't know whether he was trying to be nice or whether he was trying to find out if I was crazy. When I enlisted I weighed 160 pounds...when I was discharged I weighed 110 pounds. The military wanted to keep me in a hospital for observation, but

I declined to let them do so. All I wanted was to get out of the service, no matter what would occur in the future.

I had no idea what lay ahead for me, but I knew what I had been through and I wanted no more of that kind of life.

I collected my pay and went home with an uncle who came to pick me up. I stayed with this uncle and his family until I started college. They were very nice to me and placed no restrictions on my actions. This allowed me to assume my own responsibilities and adjust to a new lifestyle.

CHAPTER 17

Civilian Life: 1946-1952

I WAS released from the Army in 1946 and though I did not know it at the time, I was placed in the Reserves which was later called to my attention.

I did not have much direction or ambition. All I knew was that the war was over and I was out of the Army. My grandparents had deposited the money I had sent home which amounted to about $5,000, a small fortune to me.

No new civilian cars had been manufactured during the war years, but I was able to purchase a used 1940 Pontiac for $2,000. I had no job, no skill, little ambition but I had a desire to live life to its fullest. I wanted to live every minute as if it were to be my last. I was trying to make up for what I thought was lost time spent in the war. It took me a long time to realize that you can never recover the past.

I watched my bank account dwindle on wine, women, and song. Everywhere I went (as with most other WW II veterans) I was treated as a hero. People bought me drinks, invited me to dinner, and generally accepted me, which was very gratifying. We had won this war unconditionally and there was a great sense of pride in being an American serviceman.

A girl I was dating at that time was going to college and she encouraged me to enroll in the college she was attending, which I did. The GI Bill was in effect which abetted my now meager savings. I spent a year in college until the girl I was dating got serious and wanted to get married.

I was scared of accepting the responsibilities of marriage. Perhaps because of the failure of my parent's marriage, or I had not yet experienced enough of the "good timing." The college I was attending was in Indiana, so to relieve the marriage pressure I decided to visit my father in Tennessee during the summer break. I thought if I was away from this girl for some time, I could sort out my feelings and come to some sort of decision.

My father was now living in Oak Ridge and was an electrical supervisor in the Atomic Program. When not drinking he was considered a genius in the electrical field. During the war he had installed and hooked up the electrical generators on most of the Tennessee Valley Authority (TVA) dams. Because of this he was one of the first people to be selected for the Atomic Program at Oak Ridge.

He still owned the acres which, by now, were cleared so it could be called a farm. He made good money in the electrical field which he lost in the agricultural field.

While visiting Tennessee I met a girl whose father owned a farm close to my dad's. Her father also owned a tourist court, restaurant and beer joint. This girl worked after school in the restaurant while she was a senior in high school. Her father and mine had become good friends during the war years, so he tolerated my spending time drinking beer in his bar.

He was a big, big man with a "tush hog" reputation. A "tush hog" refers to a wild boar with big tusks, the meanest critter in the world. The real reason I spent time there was not to drink beer but to get a date with his daughter. After spending a few weeks in Tennessee I reached the conclusion that I was not yet ready to go back to Indiana and get married.

I will now name the girl which I had the crush on in Tennessee, as she became my wife a couple of years later. Her name is Lois and we have been married 47 years. Our first days of courtship were very rocky. Her father was a very stern disciplinarian and he had never permitted her to date. He would take her to dances and she could be friends with the boys, but that was all. Occasionally he even let her ride home from a dance with a boy while he followed them in his car.

Lois' mother and father were separated which left the responsibility of her raising and her young brother on his shoulders alone. A responsibility he assumed with great love and care. A better person I have never known. He has become a father to me, not a father-in-law.

Back to my story...I kept asking Lois for a date, but she would not say yes because she didn't want to ask her father's permission and she would not go out with me without it. I kept hanging around her restaurant making small talk and kidding her about a date. I could tell she liked me, but she would not admit it. Finally, a few days after she graduated from high school and knew she was going to college in the fall, she did ask her father to allow her to go to a skating rink with me. Her father was friendly with me but I'm not sure he trusted me. He said she could go but she had to be home by 12 p.m. We went to a town about 30 miles from her home and skated. When we started home a big storm came up which caused a considerable delay in our arrival. We arrived at her house at 12:15. Her father was standing on the porch with the light on; remember, he was a big man with a mean reputation. He came out to the car and I rolled the window down expecting the worst. All he said was, "Young lady, you just made two trips tonight, your first and your last." With that he turned and went back into the house. Lois got out of the car and quickly went into the house, I had not even kissed her good night. Years after we were married we would kid him about that evening.

My father got me a job in the laboratories at Oak Ridge which at that time was still gated and restricted to workers and their family. It was a highly secret installation and the work areas were closed to the outside public, but the social life that went on in the living area was no secret.

Oak Ridge was constructed by the government in a remote area of Tennessee. The government erected temporary houses for married workers and dormitories for the many single men and women who worked there.

For about a year I lived in one of those dormitories or more accurately stated, I changed clothing there and occasionally slept there. I worked enough to get paid so I could party. Men my age were in the minority and the women, two years younger to 10 years older were overly anxious to entertain or be entertained. Having spent almost five years in the service I thought I was enjoying the fast life. On one occasion when I met my father after an all-night party, he told me that my eyes looked like two "piss holes" in the snow.

Lois, the lady I was to marry, came to Oak Ridge to stay with her mother and attend the University of Tennessee. She no longer had to ask her father if she could date so she started going with me. At first our courtship was off again on again and all our friends made bets that the marriage would never happen, but finally I chased her till she caught me. The smartest move I ever made was to propose to this beautiful girl.

She turned my life around, she encouraged me to go back to college which I did. At the end of her first year in college she quit and went to work to support us, as we had decided that for the future it would be better for me to obtain a degree at this time. About a year after we were married Lois became pregnant with our first child, a son.

In my last year of college we moved back to Crossville where her father had a grocery store, filling station, and a feed business. We lived in the back of the store and I would drive about 60 miles each way to attend classes while she pumped gas and tended the store. Things were going pretty good in my life and I was very happy. I had a new wife, a new son, a new Ford convertible, and a new sense of responsibility. Gone were the days of fast living and I found I didn't miss the drinking, the all-night parties, or any of those things. I was really living the good life now.

I loved to hunt and fish, so one Sunday morning after working 10 to 12 hours a day when I wasn't going to school, I went fishing in a neighbor's pond. When I returned to the store where I had left Lois working, my father-in-law was there. He obviously did not approve of my going fishing and leaving Lois to tend the store and he made it plain when he told me, "You will never amount to a damn, all you want to do is hunt and fish." Well, I told him, "I'll be the best hunter and fisherman you will ever know." Lois and I still remind him of what he said, because after I finally retired from the Army I obtained a charter captain's license and fished regularly as a part of my recovery therapy. My father-in-law even became one of my best fishing companions.

At the end of the winter quarter at the University of Tennessee in 1952, I received a notice to report to Fort Jackson,

South Carolina, for two years of active duty. In other words, I was drafted. It seemed there was a terrible shortage of infantry offices in Korea, and even though I did not know it when I accepted the Presidential Appointment, I was automatically placed in the Reserves subject to recall in any emergency.

I was bitter about being called back to the Army, as I thought I had done enough for my country already, and I did not want to give up the good life I was enjoying. If it had not been for my father-in-law, I would have tried everything to get out of returning to the Army. This wise old gentleman's father had been in the Spanish-American War, and had drawn a military pension which had maintained his family during the depression years.

Drawing on this experience, my father-in-law encouraged me to return to the Army and make it a career since I already had five years of service. This I did very reluctantly. I sometimes wondered if his advice about going back into the Army was not used as a ploy to get rid of me, remember he said I would never amount to a damn. I really wrote this in jest because I know of his deep affection for his children and me. He is very wise in the ways of the world and I have always sought and heeded his advice.

Recall

My bride, Lois Jean (Dykes) Walker in 1948. This picture was carried by me in Korea in 1952.

CHAPTER 18

Recalled to Active Army Duty

I ARRIVED at Fort Jackson at the same time many reserve training officers were reporting; these were new college graduates, just commissioned, with no prior service. The only experience they had was a couple of months training at a summer camp and their class work in college. Even though I had just been in college I was three or four years older than most of them and have been through WWII, but I was still only a Second Lieutenant, the same rank they held. Had I known I was in the Reserves I could have participated in a Reserve unit and been promoted to First Lieutenant or possibly Captain.

The Post had set up a two-week orientation course which was unofficially called a "charm school" by experienced officers. This school was an orientation program used to teach new officers how they were supposed to act. Fort Jackson was the home to a basic training division to which new recruits were assigned to a company. The company was usually commanded by a Captain and he was assigned two or three new Second Lieutenants whom he was to supervise while they learned how to conduct their duties as officers.

The company provided billeting, supplies, and mess facilities for the trainees assigned. They taught close order drill, how to live in a barracks, and took the trainees on long marches. The company did the administration, issued orders, and assigned all the housekeeping duties to the trainees.

The actual instruction of infantry skills was taught by an instructional committee to which I was assigned. The procedure was for a new Second Lieutenant to be assigned to one of the committee teaching units where he would observe an instructor that had memorized the lesson. This period was a waiting time where the new officer was given time to purchase his uniforms, attend "charm school," then take over parts of the instruction.

I was assigned to the machine gun instructional committee to observe another lieutenant giving instruction on the machine gun. My fourth day there the Regimental Commander and the Adjutant, a Captain who later became a very good friend, arrived at the training site. I reported to them as I was just standing around observing. To be honest I was doing nothing. The Regimental Commander told me he was going to relieve the Lieutenant instructor I had been observing and tomorrow I was to take over instruction on the machine gun. I started giving excuses about not having my uniforms, had not observed this instructor enough, and not having attended the orientation school (charm school).

The Adjutant cut me off. He said, "Cut the crap, Walker, we checked your record and we know you were a machine gunner in WWII." They then told me to take the rest of the day off and be ready to present eight hours of machine gun instruction the next day. I never did get to attended charm school, so I never have learned to <u>act</u> like an officer.

While I was on the machine gun instructional committee the Army introduced the 57-millimeter recoilless weapon. I was chosen to form a demonstration crew to prepare the instructional material and present the training to all recruits in my regiment. The division consisted of three regiments, each with its own instruction committee.

The Commanding General wished to view the instruction given on this new weapon, so he ordered all three instruction crews to give their presentations just as they would be giving it to the trainees. The other two crews had Captain instructors, so he began with them. This General was the meanest man in the Army. In private he was referred to as the "Red Snapper" among other names, but to his face it was always, "Yes sir" or "No sir."

The first Captain and crew started their instruction and when they were about halfway through the Red Snapper stopped them and told the Captain to get the hell out of his sight.

Then the second Captain and crew started their spiel. The snapper didn't even allow them to get halfway through before he stopped them and told them they couldn't teach a dog to bark.

He then turned to me and said, "Lieutenant, see if you can do better." I was scared but I never let it show. Anyway, I went through my whole presentation without being stopped. When I ended, the Red Snapper just walked off without saying a word. I thought my officer career was over before it started. When I got back to the regimental area I was told to get dressed in a Class A uniform and report to the Commanding General. I had visions of being booted out of the service and all kinds of disciplinary action. I was pretty sure of one thing

— that I would not be put on KP because officers did not do KP. Besides, the laws had been changed in the New Uniform Code of Military Justice, which prohibited placing men on KP as a form of disciplinary punishment.

When I reported to the Snapper, he asked me if I would like to be his aide. He then told me this was a voluntary assignment and I was free to accept or reject the offer.

I knew the General had a daughter and it was part of the aide's duty to escort her to military functions. I also knew the reputation of the General as a stern disciplinarian. He would have a Military Police truck follow his car and if any soldier failed to salute his car he was picked up and taken to Headquarters where the soldier's Regimental, Battalion, and Company commanders were summoned. All were then given a lecture on military courtesy and discipline. For these and other reasons I declined the offer to be his aide.

I had been told that I might be given command of a company, so I told the General I wanted to be a career officer and the chance for command duty was very important to me. He accepted my refusal by saying, "Lieutenant, your instructional crew did a nice job." Much later in my career I realized that the stern disciplinary image he displayed was meant to instill discipline in his troops.

CHAPTER 19

First Command

A FEW days after my interview with the General, I was given command of a company and what a company it was. The draft had been in effect for several years when the government decided to try an experiment. They were reevaluating men who had been disqualified the first time through the induction process. The government was conducting an experiment to determine if any of those rejected men could be salvaged and trained as infantrymen. I was given men who could not read or write. Some claimed to be conscientious objectors, others had physical problems, real or imagined. And others who had fallen behind in their training in different units.

The latter group consisted mostly of men with previous disciplinary problems in their training units. I believe I was given command of this group because of my previous enlisted service since I was the only Second Lieutenant Company Commander in the Division. Maybe I was chosen because the

Adjutant whom I spoke of earlier had been an enlisted man and recalled to service as a Captain. In retrospect, he became my mentor and helped me get through my first command experience. The training cycle lasted eight weeks which was the tenure of my command period.

I will briefly recount some of the problems with which I was confronted in my first command. The first one that comes to mind concerned a man who had been missing for three days. I was about to charge him with being absent without leave when we found him in a tree top outside a little regimental officers' club. He claimed to be a birdwatcher and he had never left the post. The outcome of this case resulted in this man getting out of the service on a section 8, which was a psycho discharge.

Another man would stand up in his cot at night and urinate in his bed then sleep in it. My First Sergeant, a tough old soldier, arranged a GI bath for him which cured this man of that bad habit and who later became a reasonably good soldier. GI baths at this time were frowned on in the service and I did not know it had taken place until later. Had I known it was to take place I'm not sure I would not have approved of it; anyway, it cured one soldier of a sick behavior, at least while under my command.

It was mandated that on long hikes all men and their equipment would complete these marches. If a soldier was unable to carry his equipment the other men in the company would assist him. We had one soldier that was so weak he could hardly walk. He only weighed about 90 pounds. I personally carried this soldier on my back half the march. When we returned to the company area I carried him on to the dispensary where I told the doctor I didn't care what he did with him, but he was not allowed to return to my company. This man was given a physical discharge.

Another man who was a malingerer would constantly fall out on the marches. He would lie down and cry like a baby complaining that he just could not finish the hike. I pulled this trainee to his feet and told him he was to precede the company as the point man. I then had the company fix bayonets, and I told him that if we caught him I would have the company use him for bayonet drill. The entire company double-timed behind him for almost a mile with fixed bayonets. Each time after that incident this man led the company on all hikes. I think he believed I would have used him for bayonet practice, and I'm not so sure I wouldn't have. Anyway, he finished basic training.

I had another soldier who proclaimed himself to be a preacher, so I gave him a small room where he could hold religious services. I figured any help I could get in this area would be beneficial, which it was. He recruited about 10 followers who gave me little trouble.

The basic training company was larger than a normal infantry company, and everything was done according to a strict time schedule that went like this: Reveille was 4 a.m.: wash, shave, dress, make up bunk until 6 a.m.: breakfast. 6 a.m. – 7:30 a.m.: fall in for class. 7:30 a.m.: march to class and start instruction at 8 a.m. Instruction: 8 a.m. to noon. Lunch: noon to 1:30 p.m. March to class: 1:30 – 2 p.m. Instruction: 2 p.m. to 4 p.m. Free time: 4 p.m. to 5 p.m. Evening Chow: 6 p.m. – 7:30 p.m. Free time: 7:30 p.m. to 10 p.m. Lights out: 10 p.m. A 10-minute break was given each hour. My reason for giving this schedule is so the reader understands one of the problems it caused and how I managed to solve this problem.

The problem had to do with the tight schedule for feeding. The first 50 or 60 men in the chow line would have their mess trays filled with more than they could eat. The chow would run low and the last men would come up short of food. At first, because of the tight feeding schedule the men were not allowed second helpings. I altered this by allowing the men to have second helpings, but they had to eat everything they took on their trays.

I started serving on the mess line and issued an order that no one would take more than they could eat, and if they did they had to carry the excess out in their pocket. This seemed to work very well, because I monitored the garbage can where they could only deposit inedible garbage such as bones or peelings. One day I was watching the throw away garbage when a young soldier came by with a big gob of macaroni and cheese on his plate.

When I told him to put it in his pocket he started crying and said, "Lieutenant, it just isn't fit to eat." I took his spoon scooped it full off his tray and put it in my mouth, then spit it out and allowed him to dump it in the garbage. In the meantime, a line had formed at the garbage can. The men all had macaroni on their trays. Men were still coming through the line receiving their food. I grabbed my mess Sergeant and made him take the big pan of macaroni off the serving line. I had him put the macaroni in the refrigerator. Next, I ordered him to take all the food that could be prepared quickly and replace the macaroni to feed the rest of the company.

After the meal, I ordered the mess Sergeant and all the cooks to eat the macaroni. I would not allow them any other food in the mess hall until it was all eaten. They ate that macaroni for about three days, when I relented a little and let them eat other food providing they had a little macaroni for dessert. Needless to say, the quality of food preparation improved in my mess hall and we were awarded a "Best Mess" citation. For the Thanksgiving meal the Regimental Commander ate with my company. I also garnered the nickname "Little

Jesus." Like the "Red Snapper," I was never called by my nickname directly, but word gets around.

I mentioned the 10-minute break we had every hour because I was having so many disputes among my trainees. I used these breaks to settle these arguments. I acquired a set of boxing gloves from Special Services which I carried on my company guidon. During each break my company formed a circle and two disputants put the gloves on and had a little boxing match. It not only settled a lot of arguments, but it made many of my trainees get in good physical condition. At the end of the training cycle the post held a boxing tournament and my company had the most entrants and the winners. No one was seriously hurt during these fights and after the first few matches we had men volunteering to participate.

The policing of the Regimental Headquarters area was rotated among the training companies. Policing consisted of picking up cigarette butts, candy wrappers, and other debris thrown there by Headquarters cadre, who were mostly clerks for administrative personnel. When my company's turn came for policing the area, I sent the whole company to do the job before training started. Before the first break at 9 o'clock I received a call to send some men to police the Headquarters area. So, I sent a Lieutenant and one platoon to do the job during the break.

Just before the 10 o'clock break I received another call from Headquarters telling me the area had not been policed. This time I formed my whole company and marched them to the Regimental Headquarters area. I spread them out, made them get down on their hands and knees and we crawled through the area. Headquarters area was located on the main street of the post. I was on my hands and knees leading them when out came a Major, the S3 (training officer) and my captain friend, the Adjutant. The Major started barking at me asking what the hell I thought I was doing, to which I replied, "Sir, we have policed this area three times and it seems my men can't see very well. So, we are getting down close to the work in order to do a thorough job." The Major made some remark about my being a wise ass and then asked what if the General "Red Snapper" came by. To this I replied, "Sir, I would tell him the same thing I told you." The Major just turned and walked away. The Captain, with a grin, told me to return my company to the training area.

I never heard any more of this instance until the end of the training cycle when I was due to depart for Fort Benning, Georgia. The Adjutant and I were scheduled to leave, and a little departure party was held at our Regimental Officers Club. During the drinking and joking the Regimental Commander made a little speech. While he was speaking he pointed at me and said, "You may not be the best officer I ever knew, but you're sure as hell the best 'Police Officer' I ever want to know."

CHAPTER 20

Korea

I WAS scheduled to leave Fort Jackson to attend the Basic Infantry Officers Course before departing to Korea. This was my first formal infantry officer training. Nothing earthshaking occurred during this period except my Fort Jackson captain friend, his wife, my wife, and I played bridge all night. One night while we were playing, a tornado leveled several blocks of buildings near the fort in Columbus, Georgia. We really never knew it happened until the next day when we drove to the area and observed the damage. The area looked like a huge bulldozer at leveled a three-block area of homes nearly a mile long. It's very difficult to describe the powers of Mother Nature.

My wife soon returned to Tennessee and I was flown to Seattle, Washington, where I boarded a ship for Sasebo, Japan. While at Seattle, 60 soldiers were brought aboard the ship in chains. They had been convicted of minor infractions of military law and they were told if they volunteered for front line infantry duty in Korea their convictions would be commuted. Some of these soldiers had been in the company I had trained at Fort Jackson. The soldiers were all placed in one hold of the ship. I, being the only infantry officer aboard, was put in charge of these soldiers.

Only one incident occurred during the voyage. One young soldier refused latrine (head to the Navy) cleaning duty. My second in command was a big, mean, tough Sergeant with many years of service. When he informed me of the refusal to work we proceeded immediately to the latrine. I had everyone leave the latrine except my Sergeant, the man refusing to do the cleaning, and myself. When everyone had left, the Sergeant closed and latched the steel door. I began to try to reason with the young soldier, but he didn't want to listen. He took a swing at me which missed. I then grabbed him and shoved his head in a commode and told him I would flush him out into the ocean. I scared the hell out of him and he was begging me to let him up, which I did. He became quite docile and was ready to listen to reason. Because of his behavior I made him permanent latrine orderly for the remainder of the trip. I put him in charge and told him the latrine would be clean enough to eat off the floor, and if it was not up to par, that would be where he would eat his meals. All the time this "training exercise" was taking place my Sergeant never witnessed a thing, or so he said. It's surprising what occurs sometimes when you place a person in a position of some responsibility.

Every day thereafter, until the ship's arrival at Sasebo, Japan, that particular hold at the cleanest latrine on the ship.

Later in Korea this soldier was assigned to my company where I promoted him and made him a squad leader. He became a good soldier so maybe my latrine training lesson channeled his leadership in a useful direction.

From Sasebo, Japan, I was flown to Pusan, Korea as a replacement. The day I arrived I was assigned to a cavalry division which had been converted to a regular infantry division. The division had retained the customs, traditions, and colors of the old Calvary even after its conversion to infantry. This was one of the oldest divisions that had been on continuous active duty in the U.S. Army.

Four months after I was assigned to the unit, the peace talks began. This unit had suffered severe casualties and was being withdrawn from Korea. It was being shipped to Hokkaido, Japan. There, it was to be reorganized and trained to reenter Korea in the event the peace talks failed.

Hokkaido is the most northern island of Japan and the closest to Korea. On a clear day you can see Russian Siberia from Hokkaido. I tell you this, so you can get a realization of how cold the climate is. At time Hokkaido was a remote, desolate area, very sparsely inhabited. It was called the Japanese frontier. There were few trees and the soil was black lava sand. The main source of food for the natives was fish and squid. When the squid was dried on racks in the sun the stench was horrible.

The division moved into a U.S. built air base that had been vacated by the Air Force. The base could only accommodate two of the regiments, so another temporary camp was built for the regiment to which I was assigned. There was a distance of five miles between the base and the temporary camp.

The regiment to which I was assigned went on maneuvers immediately upon reaching Hokkaido. I had been given command of a company and while in the field in this cold, wet environment, I discovered how poorly my company was equipped. I had been issued a sleeping bag that had not been cleaned and I caught scabies. To get rid of them I had to take baths in kerosene every day for a week. I can tell you, kerosene baths are most unpleasant; they burn and smell terrible. I found my men walking around in low quarter shoes and some of these shoes had holes in the soles. These men had not been issued boots in the replacement centers they came through.

The day I discovered these shortages, my Sergeant and I took a truck to Sasebo where the Quartermaster depot was located. I entered the depot where I commanded the Sergeant in charge to load my truck with boots and sleeping bags. He wanted a written requisition and tried to refuse my order. I had no requisition and I did not have time to get one and having seen my man in such a desperate condition I was very angry. So I emphasized my order by pulling my pistol. When he saw I meant business he had my truck loaded with boots and sleeping bags.

During the time it took me to return to my company, all hell had broken loose. The Quartermaster Colonel had called the Regimental Commander who had called my Battalion Commander who was waiting for me to arrive.

The Battalion Commander whose nickname was "The Bull" was so mad he could hardly roar as was his custom, but he let me explain what I had done and why I had done it. The Bull was the best commander I ever had, and he really went to bat for me. When he realized how badly equipped the Battalion was he became furious and was my strongest ally.

I suddenly found myself a newly promoted First Lieutenant in direct conflict with several colonels. I could have faced a court-martial, had The Bull not interceded on my behalf. I was guilty of "doing it my way instead of the "Army way."

The Bull was the most respected officer I have ever known. He was commissioned in the National Guard and had served in WWII, then recalled for Korea. This man knew every man in his Battalion by their first name and was highly respected by his superiors and subordinates. He was a very stern disciplinarian tempered with great compassion.

This incident resulted in the complete Regiment's supply status being inspected, and we were issued much new equipment. What started as a controversy ended in a positive improvement and I never had to take a bath in kerosene again.

For the first two months on Hokkaido, we practiced invasion tactics. We would take LCIs (landing craft infantry) out to a large ship anchored two or three miles offshore, and then climb cargo nets hung over the side to board the ships. Once aboard the ship we would climb down the nets to the LCIs where we would return to invade the beach. These were the tactics we would be forced to use if we had to reenter Korea.

The first time we attempted this maneuver I was assigned to take some men to the ship which was anchored about five miles offshore. They were to help the Navy with KP duty. The Navy was supposed to feed the battalion when it boarded the ships. The seas built up and the soldiers I was escorting all got seasick. They were vomiting and the bilge water in the bottom of the LCI smelled like hell. I held one man up to keep him from drowning in the stinking stuff. When we got to the ship I had to lash some of the men and have them pulled on board, as they were too sick to climb the cargo nets.

I returned to the beach and my Sergeant and I were told to report to the Division Commander, who was a General. When I reported, the General asked me how it was out there. During the time I was telling him it was very rough, the Sergeant vomited on his feet. This emphasized the point for the General and he immediately canceled the exercise. The poor soldiers that had been taken out to the ship to help with KP had to stay aboard for three days. I don't believe they did any KP, as they were too sick.

Soon after the division was on Hokkaido we got a new Commanding General, who else but the Red Snapper. He was inspecting my battalion's training in the field one day when I had to report to him. He recognized me and asked what job I was doing. I told him I was a Company Commander.

He then said he was looking for an aide and asked if I could recommend one. I told him I did not know of anyone since most of the Lieutenants were new replacements and I knew very little about them. A week after this encounter an order came down from Division requiring all companies to be commanded by a Captain. Whether this order was aimed at me I will never know, but I was relieved of my command by a Captain. The Bull then selected me to be his S4, the Battalion Supply Officer, which also called for a Captain. He did this partly because of the boot ordeal, and by assigning me to another Captain's position it certainly helped my career pattern.

The controversy created by the boot incident followed me throughout the rest of my career as an officer. When The Bull left the Battalion for another assignment, he filled out an efficiency report on me giving the highest rating possible. This report was sent along with a letter of commendation to my career management branch in Washington, DC which caused me to be categorized as a logistician.

I would much rather have been chosen for the command field, but logisticians are considered a rare breed. They must be action people with a great deal of flexibility that can do anything necessary to make a Commander's decisions work.

In reality a supply person has to be a trader, a compromiser, a mover, and in nice terms a purloiner of government property. I never really stole any government property, but I have relocated and reallocated much property to places where it could be utilized for its intended purpose instead of being stored on a shelf or holed up in some depot.

As the peace talks continued, officers and NCOs were permitted to have their families join them in Japan on a point system. Those having been separated the longest received first choice of housing. There were three categories of housing: limited housing on Hokkaido at Sapporo, intermediate housing on Honshu at Hachioji, and private rental in Chitose on Hokkaido, if it was approved by a Regimental Commander.

CHAPTER 21

Chitose Private Rental

PERSONS WHO had been overseas the longest could have their dependents brought directly to Hokkaido until all housing units were filled. There were few of these units available. The person with the next longevity could bring their family to Hachioji where the servicemen could get limited leave time once or twice a month. They would be flown to Honshu, the main Japanese island, to spend three days with their family. These trips were called honeymoon flights where families could be reunited for a few days. The third category, private rental, was the one I took advantage of.

I located a Japanese carpenter that owned a rice paddy on the outskirts of Chitose and negotiated a deal with him to build a house. Americans could not own land in Japan at that time, but I drew up a contract on a piece of paper whereby he was to build a house that I could occupy as long as I was in the area. Papa sans, as I called him, could not speak English and I did not speak Japanese. The paper we both signed wasn't worth the cost of the paper it was written on; however, we formed a mutually beneficial relationship. I was allowed to live in the house as long as I was stationed on Hokkaido. When I left, Papa sans could move into the home I had paid him to build.

I was able to get the post engineer at Chitose camp to dump the cinders from the heating plant in Papa sans' rice paddy. This was to the benefit of the Army because they had been paying a fee to get rid of the cinders and they could dump them in the paddy free.

My agreement with Papa sans, was for $500, $250 to start and the rest when the house was finished. Papa sans, with my help, constructed a two-story house on the cinder-filled rice paddy. It took almost two months to build what was considered a mansion in Chitose, but would have been called a shack in the U.S.

The house had oiled paper sliding doors with no locks. The water line between the two Army camps ran alongside the rice paddy and I was able to tap into it so we had potable water piped into the house. I had Papa sans build a box-like sink of plywood and lined with galvanized tin. To form a grease trap we had two elbow pipe fittings welded together. This was done to keep the cold air and odors from coming up the drain. The drain was left open outside and just ran into the cinders that were used to fill the rice paddy.

We managed to find a commode with a gravity fed tank which allowed us to flush it into a canal that ran by the house. We mounted a 50-gallon oil drum in the attic to provide water for a shower. There was not enough pressure from the water line to get water up to the barrel, so we put a hand-operated gasoline pipe in the line.

When we wanted to shower the first had to pump the barrel full. We heated the shower water with a charcoal hibachi placed under the barrel. This was slow, but it worked. I found a kerosene stove for cooking and was issued a space heater by the Army in which we burned diesel fuel for heat.

When the house was nearly completed, two other officers had Papa sans build each of them a house and the rice paddy became known as POL Alley (POL abbreviation for petroleum oil lubricant). As crude as it was I was able to get Army approval to use it for housing and send for my wife and son who were allowed to join me on a space available basis. About a month after the house was completed my family received orders to join me.

A Japanese village had grown up around and between the airbase and the temporary camp, which were about five miles apart. The village consisted mostly of bars where prostitutes plied their trade.

The prostitutes were chattel to the bar owners who had purchased them from their parents. It was a common practice for poor peasant farmers in Japan to sell their girl children at a young age, and some of them were forced into prostitution by the age of 12. The Japanese girls of that era were taught that their only purpose in life was to work and please the men who were their masters.

The bar owners usually called Papa sans, could make the girls do anything they wanted them to. Many girls were made to stand outside the bars in the bitter cold until they had turned enough tricks to satisfy their owner's demands.

I saw girls standing naked in bar windows with rotating red, white, and blue lights on them. For this reason, Chitose became known as the "Sex Circus of the Orient."

It was alleged that there were 5,000 registered prostitutes in Chitose and that on a pay day this number increased by 2,000 unregistered prostitutes.

Lois and Skeet at our private rental in POI Alley on Hokaido, Japan

Little did my wife and son realize the hardships and terrible conditions they were to encounter. Hokkaido was dry, cold, and windy where 48 inches of snow fell in one night. The only means of transportation was to catch a military bus that ran between the two camps or walk. This was a hell of a place to ask a "little ole southern gal" who had lived a sheltered but austere life, to come and bring our four-year-old son.

When Lois and Skeet (my son's nickname) were to arrive by ship in Yokohama, I was given seven days leave to meet them. It was a two-day train ride each way from Hokkaido to Yokohama. The train crossed the straights of Hokkaido in the hold of a large ferryboat. The trip in the boat took about six hours. I arrived in Yokohama where I spent the night in a hotel.

Around midnight I was awakened by people screaming and the building shaking. It was the first earthquake I had ever experienced. People were running to get out of the building. I arose and looked out the window where I saw the narrow streets packed with people. I thought to myself, if the hotel fell it would fill the street with rubble and crush all those people, so I just lay back in bed where I resigned myself to fate. The earthquake caused little damage to the buildings, as Japanese buildings are not built on a rigid style but are constructed to give and move with an earthquake. After this earthquake, I experienced many smaller ones called tremors. A tidal wave caused by the earthquake resulted in the ship which my family was aboard being delayed 24 hours before docking. Words cannot express the joy I felt being reunited with my wife and son after so many months.

We spent a day and night in Tokyo and then started the train ride to Hokkaido. The train was sooty and dirty, but in spite of it all it was the happiest train trip I ever made. Just being with Lois and Skeet was the greatest feeling I ever had. No matter how bad the conditions were, we were together.

The next day after we arrived in Chitose, I had to leave my family for maneuvers in the field where my unit was on alert to return to Korea. Three days after we had been in the field it started to snow, so the Bull told me to take time off and go check on my family. I took a weasel (a tracked vehicle) that could travel over snow and went to our home.

When I arrived at our shack, I was met by Papa sans who stated, "Oksan Toxsan pissed off" translated means, "wife very much angry." Lois came out of the house and began to cry.

The space heater had exploded when she tried to light it. Soot was everywhere, and she was covered with it. Lois had wrapped Skeet in blankets and had taken him to another private rental in the area where a chaplain's wife was living.

The water line had frozen where it crossed a dirty stinking little ditch. While she was trying to pour hot water over the line to thaw it, she slipped and slid into the stinking ditch. She really was PISSED OFF in capital letters. She stated that she was going home to Tennessee.

I asked her how she was going to get there and when I took her in my arms she never answered. Papa sans and I managed to get the space heater lit and the water pipe thawed. We retrieved our son, prepared a good dinner, and spent the night together. Lois, a very resourceful lady, never complained to me again about the harsh conditions she experienced. My wife is the most wonderful person in the world; how she has put up with all the hardships I caused her is almost beyond comprehension. It must be love.

The next morning, I left for the field where the unit stayed for 10 more days before returning to our base camp. I could now spend the nights with my family in our mansion in the dump. The peace talks drug on for over a year while we were stationed on Hokkaido.

The Calvary unit to which I belonged was the only unit in the Army that celebrated annihilation day. It was the historical unit that had been with General Custer at the battle of the Little Big Horn. Even though it had been forced to withdraw (Army terminology for retreat) in Korea, it maintained many of its proud traditions. One of those traditions involved the song, *"Gary Owen,"* which allegedly was played by the Regiment on the way to the Little Big Horn. Soldiers used *"Gary Owen"* for saying "hello" when meeting an officer or answering the phone. The first time I heard this greeting I thought the trooper said, "Carry on" and I proceeded to give him a lecture. Finally, he was able to explain the tradition of "Gary Owen" to me, which left me feeling very silly while I apologized.

The three years I spent in this Cavalry Regiment was the most memorable time in my career. While living on Hokkaido, a situation occurred that caused a congressional investigation. A soldier could register a prostitute as his "only" with the Japanese police by having his picture taken and placed with hers on an identification card kept in the police station. This action kept the Japanese police from arresting the women for prostitution. I have no idea how many of these registrations there were, but the Army was afraid that the Russians would learn of this procedure and exploit it for propaganda purposes, thus creating an international incident. It could have been alleged that these girls were purchased by our soldiers. Congress determined the division should not be stationed in such a lewd sexual environment, so a "secret" move was ordered to the main island of Japan. A secret move in the Army has always been a joke. If the military wanted an enemy to know something, all they had to do was classify what they wanted to be known as "Top Secret," and you could bet on it being learned by whoever wasn't supposed to know about it.

CHAPTER 22

Move to Honshu

MY WIFE and I had been living on Hokkaido in our private rental waiting for enough time to become eligible for government housing on Honshu (the main Japanese island). About the time we became eligible for housing there, the congressional investigation detailed the division to move to Honshu.

Many American soldiers had married Japanese girls at this time and the Commanding General's wife had started an orientation program for these Japanese brides. The General's wife enlisted Lois to help in this project. She and other officers' wives used our private rental to show the Japanese brides what could be done to establish an American style home for their husbands. The American wives hosted teas and other social functions for the Japanese brides which created a bond among them. This helped relieve their tensions and boredom. Lois was very involved in this project and became a very good friend to many of the senior commanders' wives.

Before the secret move of the division started, most of the onlies had already preceded the units to the main island and established themselves outside the new post. The military could never keep a secret.

I was now eligible for government housing, but the post to which we were moving was being used as interim housing for dependents waiting to move and join their sponsor on Hokkaido. All the quarters were occupied by less eligible families than mine, but the Army could not put them out as there was no place to put them.

My Regiment was the first deployed to the new location and I was forced to leave my family on Hokkaido in the private rental alone, except for the three other families that had built in POL Alley. Two regiments remained on Hokkaido which provided some comfort and assistance to these families. When the second Regiment deployed, some government housing became available on the airbase at Camp Chitose and private rental families were moved into them.

My family was moved into the now vacated General's house which had been built for an Air Force General at an exorbitant price to the government. I heard he had been court-martialed because of the high cost, anyway it was like a castle. Lois and I described this move as going from the "Shit House" to the "Big House." The division started a reverse honeymoon flight where I flew back to Hokkaido for three days once a month.

The Army obtained some an unoccupied government housing on an air base near the new division post where my family and others could live pending availability of on-post housing. We were assigned one of these interim quarters and prepared to move.

Lois was issued orders that would let her accompany the last Regiment moving south by train. On my last honeymoon flight, we decided Lois and Skeet should fly south on a Japanese airline at our expense. The reason being that they had both gotten seasick when riding the ferryboat across the straights of Hokkaido.

This turned out to be the best decision we ever made. The ferryboat on which they were scheduled to travel was caught in a typhoon and overturned, trapping the soldiers and many dependents in the train. Several hundred lives were lost in this accident including many of our friends. Note I do not call them "close friends," as I learned not to make close friends in the combat services because their loss could be very traumatic. The loss of lives in this tragedy affected the entire Division and the sorrow was felt by all the survivors the remainder of time I was in the Division. I can still feel the pain of that loss.

About three months before our unit moved south, I had appropriated 15 gallons of red paint which we were going to use to paint the latrines in the Quonset huts the companies used for barracks on Hokkaido. They were cold, rusty, dirty temporary buildings which the Quartermaster refused to do anything to repair or refurbish them. The Bull told me to try to get some paint to at least make them look cleaner and that the troops could paint them. I don't believe he knew of the impending move.

I, being the S4, was in charge of obtaining supplies. So, having been involved in the resupply of the boots issue I took a little different approach. This time my Sergeant and I located some paint near the post firefighting station. We mounted a midnight requisition expedition. We relocated 20 gallons of red paint and stashed it in the quarters of a major who was the S3 (Tactical Operations Officer). He was the senior officer on the Battalion staff. The S3 was a single officer whose wife had been in Japan and was sent home as an undesirable dependent.

He was in the process of obtaining a divorce, a tragedy caused from a long separation in Korea.

The major had been outstanding in combat in Korea but had developed an "I don't give a damn attitude" and did not spend much time in his quarters. Some other unit or person stole the paint from the Major's quarters. The Bull had been on a seven-day leave and when he returned he wanted to start painting the latrines. I had to tell him someone had stolen the paint out of the Major's quarters. He roared at me in front of the Major and wanted to know why I had given the paint to the S3. The Bull said, "You know Bod (the S3) can't even take care of his cigarettes, so why would you give him the paint?"

I had called Lois and told her about the paint ordeal and told her I was bringing The Bull home to lunch, which I did. When the Bull and I got to my shack, Lois mentioned something about the paint that was missing. The Bull asked her how she knew about it, and she told him she heard him roaring at Bod about it all the way from the camp. He really got a big laugh out of that, he was a great person as well as my commanding officer. I can still see him in my mind's eye a big burly, gruff man with a heart of gold.

A week before my unit moved south The Bull rotated home, but he left me with a legacy of being the best S4 in the division. This legacy remained with me throughout my military career. I never forgotten The Bull. He taught me more about leadership than any other officer I have ever known.

When we moved south to Honshu my family and I were given interim quarters on an airbase about 30 miles from my post. My private car arrived from the States where it had been stored for a year, so I was able to commute from my quarters to the post. We lived on the airbase for three months until quarters became available at my duty station and then we moved again. Three moves in six months would seem bad to some people but it did not bother Lois, Skeet, nor I.

The automobile I had at that time was a 1950 Ford four-door sedan. I sold this car about four months before I was to return home to a very rich, old Japanese man. This rich Japanese could not drive, so he brought a young Japanese man to drive for him. The young man tried to lift the car by the bumper and when he could not, he let go and said, "hebby, hebby." I think he meant "heavy, heavy." Anyway, they paid me several hundred dollars more for the car than I had paid for it when I purchased it. They paid me in Japanese yen which I sold to the soldiers on paydays in exchange for American script. It took the entire four months I had remaining to exchange the yen for script.

Just one Regiment was stationed on this base, the other two Regiments and Division Headquarters were moved farther south on Honshu.

The division was undergoing an almost complete turnover in personnel. I had one Battalion Commander for three months who moved up as the Regimental Executive Officer. The new Regimental Commander had never seen combat which caused a complete change of attitude in the Regiment. It has been said a command reflects the commander. It seemed the Regiment went from a combat ready force to a garrison ready spit and polish unit.

My third Battalion Commander arrived with his wife and daughter fresh from the States. He was a good man that had progressed through the officers ranks to Lieutenant Colonel without overseas service. Even though I was only a First Lieutenant, he relied on me as a stabilizing influence because of my longevity.

With almost a complete change of personnel we went into an extensive training program. We had a maneuver area approximately 60 miles from our base, in a remote area of Honshu. This area was only large enough for one battalion to train at a time. The Regiment would send one battalion to the area for 30 days training on a rotational basis.

There were no facilities at the maneuver area, so each unit would set up, take down and live with their authorized field equipment. In the winter it snowed, the wind blew, and it was cold as hell living in tents.

The Regiment even tried over-snow training. We were issued skis, and weasels (over snow vehicles we had used on Hokkaido). The over-snow training created a great number of broken bones among the troops as most of them had never even seen skis much less tried to use them. The officers, including myself, had never skied and we were the instructors, a case of the blind leading the blind. The over-snow training was discontinued after the first month because of the many injuries we sustained.

Every time it came my Battalion's turn to go to the maneuver area, my Commander would inquire if I had packed all of my personal equipment. The personal equipment of which he spoke was a shotgun and shells. The reason for taking this equipment was to hunt pheasants which were plentiful in the maneuver area. When we would get established in the area, the CO would have me take time off and go hunting, which I really enjoyed. I would bring the pheasants I had killed back to the mess area where the cooks would prepare them.

Lois and I moved had into quarters next door to the Regimental pilot who on a weekly basis would fly orders and mail to the unit in the field. Lois would bake pecan pies which the pilot, along with booze and cigars the CO ordered, would fly to our unit in the field.

Once a week the officers would go down to a little nearby village and use the Japanese public bath. After bathing, we would put on clean clothes and return to our camp in the field where we would hold a formal dinner party consisting of pheasant with the vegetables topped off with pecan pie, brandy and a cigar. We even did this when the General visited us on an inspection trip. He was quite impressed with how we could improvise under such harsh living conditions. We would have had pheasant under glass, but we had no glass.

The troops were also allowed to visit the Japanese bathhouse on a rotating basis. This bathhouse was used by both Japanese women and men at the same time without segregation. Nudity was not perceived in Japan as it was viewed in the U.S. However, on our second trip to the area the Japanese had installed a screen in the bathhouse, so that the women bathed on one side and the men on the other. This came about because the GIs were using the baths and staring at the naked women.

When the unit was not on field maneuvers our time was used on spit and polish duty, parades, inspections, housekeeping duty, physical training, and preparing for the next field trip.

CHAPTER 23

Change of Command

THE NEW Regimental Commander assigned was a Bird Colonel who had never seen combat and should have never been given command of an infantry unit. This Colonel wanted a boxing team and I was designated as an additional duty to form, train, and supervise the team. I chose individuals from each of the battalions and housed them together. These troopers only duty was to train and box. They were given extra privileges which lead to disciplinary problems. Battalion and Company Commanders were very upset over giving up control of their men and seeing them given special privileges which caused a morale problem. However, none of the commanders went to the Regimental Commander to complain. They heaped all their complaints on me. The team fought and won the Divisional Championship and was turned over to the Division Headquarters to train for the Far Eastern Championship. This relieved me of this bad situation.

The Regimental CO had a golf course constructed on the post and had each of the Battalion Commanders appoint a man from their unit to act as a representative. When my Battalion's CO returned from the meeting, he told me that I had been appointed as the Battalion representative. I told him I knew nothing about golf and had never been on a golf course, to which he replied, "You have been working too hard, so tomorrow afternoon you, the Regimental Commander, the Chaplain, and I are going to play golf at the grand opening." I tried to protest by saying, "Sir, I don't have any golf clubs." He cut me short by saying, "I've taken care of that, I just bought a new set and I'm selling you my old ones."

The next afternoon I played my first game of golf. I was a First Lieutenant playing with a full Colonel and two Lieutenant Colonels. I was really intimidated, but I managed to get through the match with the help of my CO and the Chaplain. Later, Lois and I played golf every chance we had, both using the old set of clubs sold to me by my CO. We both became respectable golfers and when we had the chance to play, golf became one of our favorite recreations.

An annual efficiency report was submitted to the Department of Army on each officer, or each time there was a change of their immediate commander. With all the command changes in the Battalion I was accumulating many reports. This efficiency report was endorsed by the next higher commander. In my case it was made out by my Battalion Commander, endorsed by the Regimental Commander, approved by the Division Commander, and then forwarded to the Department of Army where it became a permanent part of my record. Promotions, assignments, and retention in the service was predicated on these efficiency reports.

In the beginning, officers were not shown their efficiency reports before they were sent to the Department of Army. Just before I left Japan it was required that an officer be shown his efficiency report to assist in showing him his strengths and weaknesses. Commanders were also required to counsel the rated officer in the areas of his conduct that needed improvement. I was shown my efficiency reports but never counseled. The Regimental Commander at the time was one for whom I had very little respect. He never brought his wife to Japan (maybe she would not come). Anyway, he held a meeting on a barstool at the Officers Club every afternoon after retreat sounded. He would gather the officers without dependents and regale them with tales about the "Old Army." I seldom attended these meetings because I enjoyed spending as much time as I could with my family. He had graduated from West Point and had over 30 years of service, but not a day of combat. He made a remark as to why anyone would want their wife over here with all these Japanese ladies available. Thank goodness he was only my endorsing officer. My Battalion CO, the rating officer, had brought his wife to Japan and he did not attend the meetings either.

At this time, my wife and I were the "old timers" in the Regiment and we started a custom upon the arrival of my Battalion Commander's wife and daughter. This custom was continued for the newly arriving dependents of all officers assigned to the Battalion. The train ride from Yokohama was an overnight trip, so the passengers would arrive at our post about 9 o'clock in the morning. Lois and I would meet them, secure their luggage, and escort them to our quarters where we had prepared a great southern breakfast of country ham, eggs, red eye gravy, and grits.

We had my father-in-law send us the ham and grits from Tennessee. It was the first time most of them had ever eaten grits, but it was not to be their last. Finally, we even got the commissary to stock grits so that when our supply ran out we could replenish it. This breakfast led to great friendships among the dependents in our Battalion and we really became a close-knit family.

About four months before I was to rotate back to the States my battalion was assigned a Major as S3 (Operations Officer), who was a West Point graduate with a drinking problem. Until his family arrived he would join the Regimental CO's daily bar meeting until one or both would fall off their respective stool. Someone would help them to the BOQ (Bachelor Officer Quarters) to their beds.

The Battalion held an inspection every Saturday morning at 6 a.m., and if the units passed the inspection they had the afternoon and Sunday off. This inspection was conducted by the Battalion Commander and his staff consisting of the S1 (Adjutant), S2 (Intelligence), S3 (Operations), and S4 (Logistics). On Saturday morning the S3 failed to show up and the CO sent me to his BOQ to see why he hadn't arrived for duty. I found him still clothed, lying on his bed still drunk. I couldn't wake him, so I drug him into the shower, sat him up against the wall and turned on the shower and left him there. I reported back to the CO that the Major had overslept and would be in later. When the S3 did arrive, the inspection was over, and he accused me of trying to drown him. Soon after this he was transferred to the Regimental staff and my Battalion CO assigned me as the S3.

I was now filling a Major's position in charge of running combat field tests for all companies in the Regiment. After the company combat tests were finished, we started Battalion combat tests. When my battalion went through Battalion field tests, I was the only First Lieutenant S3 in the Division and my battalion scored the highest in the Division. The remaining Battalion S3's were all Majors.

I was scheduled to rotate to the U.S. after the Battalion combat tests ended. I was glad to be returning to the States but sorry to leave the Battalion in which I had been assigned longer than any other officer or enlisted man. For over three years I had been a Company Commander, an S4, or an S3 in one Battalion.

CHAPTER 24

Return to the U.S.

ON OUR way home, Lois, Skeet, and I had a three day layover in Tokyo where I tried to get Lois to go with me to a Japanese bath, but she never would. The Japanese families all bathed together in the nude. We sailed home on the *USS United States*, a large ocean liner. We were assigned a nice cabin on an upper deck amid ship, an excellent location for a cruise.

Twelve people were assigned to a mess table where we were served delicious food family-style. There were more dependents aboard the ship than active military people. Many of the dependents were Japanese war brides and children that were joining their husband or sponsors who had previously been returned to the US.

The first morning at sea all 12 seats at our table were occupied for breakfast. For the lunch meal there were eight seats occupied. At dinner only six people showed and some of them ate sparingly. After the first day out, I had the dining table to myself most of the time. The sea was very calm and except for gigantic swells, but at least half the passengers were seasick. I had so much to eat I gained over ten pounds.

Poor Lois, she hardly left her bunk for 14 days, the time it took to cross the Pacific. I would make Skeet go up and walk around the deck with me. I tried to get him to eat a few bites, but he never kept much food down. When we arrived in Seattle, they both looked like skeletons and I look like a butterball. Lois even got land sick for two days while we were waiting for a flight to Tennessee. She told me if she got airsick on the airplane she was going to jump out the emergency exit. I have never been able to get either of them on a ship since. When we got to Tennessee her father wanted to know what I had done to his daughter. I was home on leave for 30 days, during which time Lois and Skeet both got their appetites back.

My next assignment was to Fort Benning where I was assigned to an experimental brigade that was to devise tactics that could be used in the event of atomic warfare. The concept proposed for this unit was to have a Division Headquarters and a Trains Command for logistics, to this headquarters could be assigned one to four separate brigade headquarters. To each of the brigades would be assigned as many battalions as needed to accomplish their mission. This plan was supposed to create flexibility in structuring a force for whatever situation arose and would aid in dispersing our forces. Because of the efficiency reports given me by The Bull in Japan, I was assigned as a Brigade S4. The position called for a Major and I was still a First Lieutenant.

The Brigade S4 was in charge of all supplies for any or all units assigned. I was given one Warrant Officer and three NCOs to obtain, store, transport, issue, and maintain mountains of equipment. My predecessor, a Major, had been able to get himself reassigned and I was supposed to inventory all the equipment and sign for it. I was given 30 days to do so, at which time I would be responsible for all the equipment whether I signed for it or not.

My section and I worked day and night because there were many, many major shortages, which if I signed for them I would be expected to pay for them. This unit had been on field maneuvers the month before I arrived and had lost equipment everywhere. The Warrant Officer and the NCOs had scrounged, traded, and salvaged most all of the equipment that did not have a serial number to make up the shortages. I was being harassed by a chain-smoking Lieutenant Colonel Commander to complete the inventory and sign for the equipment. Near the end of the 30 days I told the CO I would sign for all the equipment if he would put the missing serial numbered items on a Report of Survey. A board of officers would then determine if their loss was justified. He refused to allow a board of survey because most of the items were weapons which should have been secured at all times, and he was afraid it would reflect on his efficiency report. He told me that I "would" sign for all the equipment or get out of the Army. I told him I could not afford to stay in the Army at that price. This man appeared to be a nervous psychotic that thought every officer in the unit was out to take advantage of him.

I went to the Division Trains Commander, a Bird Colonel I had known on Hokkaido, and explained my situation to him. He told me to return to the unit and say nothing. The next day the Trains Commander and all his staff came into the Brigade and conducted a thorough investigation. When they completed their investigation, they relieved the Brigade CO, put the Executive Officer in charge and informed me to turn in the

Report of Survey. After this fiasco, I felt I should either find another assignment or resign my commission.

The Executive Officer talked me out of resigning and got me an interview with Third Army headquarters in Atlanta, who at the time was looking for recruiting officers. With my combat record I was accepted as an induction officer in Knoxville, Tennessee. I spent three good years in Knoxville where I attended the University of Tennessee. While there, Lois and I purchased a new three-bedroom home for $12,500, a lot of money at that time. We purchased it under the GI Bill and lived there three months' rent free while waiting for the loan to be approved. I used to lie awake at night worrying about owing $10,000. I kept thinking I was in debt for the rest of my life and would never be able to pay for it. But when we left we were able to sell the house for a profit. This was a big help because the pay of an Army officer was very low and the pay of enlisted men even lower.

As the Induction Officer, I swore all recruits and draftees into whichever service they had chosen. While there I talked to my brother-in-law, who was attending college in Knoxville and living with us, into enlisting in the Air Force. He was going to be drafted and I told him he would have a better life in the Air Force than in the Army. After enlisting, he made a career of the Air Force and retired at the rank of Colonel after 20 years service.

At the end of three years in Knoxville, and five years as a First Lieutenant, I was promoted to Captain. With my promotion came orders to report to Fort Benning, Georgia, to attend the Infantry Officers Advanced Training Course. The Infantry School held two advanced courses, one for most Reserve officers which lasted three months. The other course was for regular Army officers and a few Reserve officers that had been selected for retention on active duty, this course lasted a year. I was selected for the year-long course since I had been selected to be retained on active duty for an indefinite period.

Ever since I returned from Japan, the military forces have been undergoing a reduction in force program. Most Reserve officers were released from service. If they had been enlisted men who held a permanent NCO rank before having their Reserve commissions activated of the Korean War (Police Action), they could revert to their NCO rank and remain service. I knew several Colonels that served in Korea who were forced to revert to the rank of Sergeant so they could finish their time in service for retirement. In many cases this was very traumatic and showed very little appreciation by our government for the tremendous service these heroes gave to their country. Those officers who were not selected for retention were released from active duty and in many cases, they faced a very uncertain future. Many had four or more years service and planned to make a career in the service. I was lucky to have been selected for an indefinite appointment, which insured my retention as an officer until I completed 20 years service for retirement.

CHAPTER 25

Infantry School

DURING MY tenure at the school at Benning, nothing much occurred to write about. It mostly consisted of theoretical tactics which I had already encountered in actuality. The only difficult part was each officer had to write a 10,000-word monograph on some military subject. I chose to write about why the Atomic Forces Concept would not work. My writings created controversy between the tacticians and logisticians. Tacticians can form great plans on paper and in theory they can move men and equipment all over a map. However, Logisticians have to be more practical because they have to physically do all the moving. Tacticians claimed the tail should not wag the dog, to which the logisticians replied, "If you cut the dog's tail off he would soon bleed to death." I would like to quote an article that was given to me later on this subject. Unfortunately, this article was unsigned so I cannot give credit to the author.

Lt. Col. "Art" Walker, Ret.

How Many Logisticians Do You Want?

"Logisticians are a sad, embittered race of men, very much in demand in time of war, who sink resentfully into obscurity in peace. They deal with facts but must work for men who merchant in theories. They work during the war because war is very much fact. They disappear in peace because, in peace, war is mostly theory. The people who merchant in theories, and who employ logisticians in war and ignore them in peace, are Generals. Logisticians hate Generals.

Generals are a happily blessed race who radiate confidence and power. They feed only on ambrosia and drink only nectar, except when they are drinking bourbon. In peace they stride confidently and can invade a world simply by sweeping their hands grandly over a map pointing their fingers decisively up terrain corridors and blocking defiles and obstacles with the side of their hands. In war, they must stride more slowly because each General has a Logistician riding on his back, and he knows that, at any moment, the Logistician may learn forward and whisper, "No, you can't do that." Generals fear Logisticians in war, and in peace, Generals try to forget Logisticians.

Romping along beside Generals are strategists and tacticians. Strategists and tacticians do not know about logisticians until they grow up to be Generals—which they usually do—although sometimes Generals will discipline errant strategists and tacticians by telling them about logisticians. This sometimes gives strategists and tacticians nightmares but deep in their hearts, they do not really believe the story—especially if the General lets them have an occasional drink of his nectar of bourbon

Sometimes a logistician gets to be a General. In such a case, he must associate with Generals, whom he hates; and on his back is a logistician whom he fears. This is why logisticians who get stars get ulcers and cannot eat their ambrosia.

I wish I could take credit for the preceding passage, even with its humorous content it really makes a point.

During the advanced course, your class standing to a large extent was determined by the results of your monograph. I do not consider myself much of a writer and I counted the words over and over, trying to reach the required quota. I only wrote 9,995 words which was five short, but evidently no one counted the words but me.

I was competing with military academy graduates from the Army, Navy, Marine Corps, and several selected foreign officers. One of the foreign officers, a major from Pakistan, had written and published a book entitled *The Khyber Rifles*. I will not disclose my final class standing, but my monograph came back marked short and to the point. It was retained in the Infantry School Library.

During the advanced course, all foreign officers were assigned an American officer sponsor to assist them with language, social customs, or any other difficulties they might encounter while residing in the U.S. The foreign officer that was assigned to me was a prince from Saudi Arabia, in addition to being a Lieutenant in the Saudi Army he had graduated from Amherst College in England, so he spoke excellent English. He was a perfect gentleman with a dark brown complexion, so when Lois and I asked a nice white lady friend of ours to accompany the three of us to dinner, she accepted. This dinner was at a fashionable restaurant in Columbus, Georgia. While dining, we received many cold stares and overheard some snide remarks which we chose to ignore.

I learned much from the Prince about the customs of Saudi Arabia, for one thing they could have multiple wives. They were not permitted to drink alcoholic beverages in their country, but he had picked up the habit of drinking scotch whiskey in England which he continued in the U.S. During the time he was in the advanced course, we became very good friends and he said when he returned home to Saudi Arabia he was going to ask for me to be assigned there. He also made the mistake of telling Lois that when I was assigned to Saudi Arabia he was going to provide me with a harem. That statement effectively eliminated my going to Saudi Arabia.

Upon completing the Infantry Officers Advance School, I was temporarily assigned to the office of the Director of Instruction at the Infantry School. My assignment was to attend parachute school for the purpose of making changes in the program primarily in the area of physical fitness requirements. Although I had already made two jumps, I still went through the school. This time no one was forced to jump. If a man froze in the door he was helped back to a seat and was allowed to land with the plane. After which he was dismissed from the school. What a difference a war makes.

I completed jump school and received orders to Korea with a bonus delay in route to Biloxi Air Base where I was to attend Air Ground School for 10 days. I had been selected to attend Air Ground School in Japan; therefore, I could compare the overseas course with the U.S.-based course. Both were outstanding, neither required tests and the only difference I saw was that in Japan I lived in a bachelor officer quarters where there were two fifths of whiskey on the dresser when I checked in. In Biloxi I had Lois with me where we lived on post in a small cabin. I had to buy my own whiskey, but having Lois with me more than made up the difference. Both air ground schools were conducted for the purpose of orienting Air Force and Army on each other's capabilities in time of war. In spite of the informality of the course it served a very worthwhile purpose.

While at Biloxi, Lois and I were able to meet another couple we knew from the Advanced Course at Benning. He was a West Point graduate and the son of a Lieutenant General. He had been assigned to Tulane University for the purpose of obtaining his Masters degree. In spite of these high qualifications he was really an outstanding officer and we were good friends. I never learned what happened to him but I am sure he attained a high rank which he deserved. The four of us had a wonderful day and night in New Orleans. We had dinner at Antion's, saw Andy Williams at the Roosevelt Hotel Ballroom, visited with Pete Fountain on Bourton Street, and had breakfast at Brennan's.

Lois returned to her father's home in Tennessee and I was flown to Korea for a second tour there. This time the war (Police Action) was over. I figured I would be assigned to an infantry division on the Demilitarized Zone, but when I arrived in Seoul, Korea, I had orders directing my assignment as S4 to the Pusan Area Command where I was to reorganize the entire supply system for the area. It seemed I was destined to be a logistician whether it was my choice or not.

CHAPTER 26

Return to Korea

*T*HIS TIME I was provided adequate help to do the logistical job, even though most of the help were Korean nationals. The biggest problem I encountered was in communication, due to the language difference. The Koreans were over anxious to please me and if I was not explicit in my orders it would sometimes lead to complications. One time I told one of them to order a load of POL and I received a call from Seoul wanting to know how many and what size poles I wanted.

In the 15 months I was in Pusan my staff and I inventoried, identified, salvaged, and catalogued supplies that resulted in an estimated $10 million savings to the U.S. government. Because of this savings I was given the additional duty as AFAK officer, which means Armed Forces Assistance to Korea. I was provided with surplus material, mostly out of date cement, rusty nails, and sheets of galvanized tin, which I could give to deserving organizations to better the relationship between the

Korean nationals and the American troops. This additional duty took up most of my time. I was expected to attend meetings with the country and city officials where I was told to which organization they preferred I give the materials. The officials were politicians who supported their friends. I had two Korean men who assisted me in checking these organizations and we found most of them had just been formed solely for the purpose of obtaining the surplus material.

My second tour in Korea was what the Army termed a hardship work, which meant my family could not join me. The separation from Lois and Skeet was the only major hardship I encountered. The Army post at Pusan contained several homes that had been built for dependents of U.S. service men on occupation duty after World War II. The dependents had been evacuated when the Korean conflict started. All the houses, except about ten, were used as officer billets. There were about ten houses used by Department of Army employees who were allowed to have dependents with them. However, I do not recall seeing any U.S. children with them as there were no schools available.

I was billeted in a house with the Deputy Area Commander, and the Area Command Motor Officer who was under the direction of the S4. He was a Major and I was a Captain occupying a Lieutenant Colonel position, but we never had any conflicts. Each of us would put $10 in our house fund each month to stock our bar. We took alternate monthly turns buying whiskey for the house. I drank Bourbon and the other two drank Scotch, so we would have two months of Scotch and one month of Bourbon. Today $30 a month would not buy very much whiskey, but in Korea in 1958 a fifth of Ballantine Scotch sold for $1.25 a fifth. When I did the buying Jack Daniels Black Label was $1.80 a fifth, the highest priced whiskey available.

The Korean people had suffered defeat by the Japanese, and during the Japanese occupation many women were forced into prostitution in order to survive. Most Koreans became thieves under the Japanese occupation, a custom that continued during this tour in Korea. The only difference being that at one time they stole from the Japanese in order to survive, now they stole from the Americans because it was easy. I had a Korean house cleaning lady that washed and ironed my clothes. When any of these items became worn I would replace them and try to give her the worn articles. She would always refuse to accept them, but when I put them under the other articles in the drawer she would steal them.

When the Pusan Area Commander first arrived in Seoul and was being driven to Pusan in a Jeep, a Korean boy on a bicycle held on to the rear tire of the Jeep pretending to be towed, but he really was loosening the lugs from the spare tire allowing the wheel to drop off. The driver and the Colonel didn't pay that much attention until they stopped and discovered the spare wheel missing. We had to put chains and padlocks on all Jeep spare wheels to keep them from being stolen and when we left a Jeep parked, we would chain the steering wheel so it would not turn, or we had to leave someone to guard it. One time the motor officer and I were duck hunting in a rice paddy, and had left our Jeep with the steering wheel chained on a dirt path a short distance from where we were hunting. We happened to look up and see ten Koreans pushing our Jeep away in a straight line as they were unable to turn the wheel. We ran after them carrying our shotguns which scared the hell out of them. They ran away and we recovered the Jeep. Had we not seen it being stolen we would have been forced to pay for it.

The motor officer and I did a lot of duck and pheasant hunting during our off-duty time. The rice paddies teemed with all species of ducks and geese. During the war, much guerrilla action had occurred which caused the Korean government to require all Korean civilians to turn in all firearms, which left them without guns for hunting. This allowed an explosion in the duck population. At first when we would shoot a duck some Korean kids would grab it and run away. After a couple of hunting trips, the kids realized we did not want the dead ducks and they became our retrievers. Near one village where we hunted, we were welcomed as heroes because we provided the whole village with ducks. They even guarded our Jeep so no one could steal it.

The method used to kill pheasants by the Koreans sometimes led to disaster. They would use a needle and hollow out a large bean into which they put strychnine. They would then place these beans along the edge of the paddies to bait the pheasants. When a pheasant came out of the hills and ate the strychnine-filled bean it would go into convulsions. Whereupon, the watching Korean would run, grab the pheasant and rip its craw out before the strychnine got into its bloodstream. Sometimes the Korean would be a little slow in getting the craw out and when this happened there would be a sick family of Koreans. When the Major and I arrived for a hunt, the whole village would turn out to welcome us. They even offered to share their food with us, which consisted of kimchi (a fermented cabbage dish laced with garlic, onion, other spices, fish or dog meat). We politely declined to partake of their hospitality. I just could not force myself to eat dog meat, although many Koreans considered it a delicacy. I think the motor officer tried it once but not twice.

The highlight of our hunts was when we killed some Siberian Snow Geese. Many of them with long legs and necks stood five feet high. We took a couple of the snow geese back to our mess where some of the nurses supervised their preparation. They were served in the mess and the officers that ate them said they were great and tasted like roast beef. I cannot vouch for the taste because I never ate any of them. After my C rations experience in North Africa, I became very particular about what I ate.

As the AFAK officer I was exposed to many Korean customs and sayings. It was said by the Koreans that, "if a girl baby lived to womanhood she had been born in the year of the full rice bowl." Meaning that if the rice crops failed the year she was born, she would be killed to prevent overpopulation and starvation. This custom was still being adhered to in the remote rural areas as late as 1958 when I was there.

Another policy, almost a custom, was the giving and taking of bribes. I was offered everything from gold to women for the surplus materials I had to give away, but let me emphasize here and now, "I never accepted a bribe in any form!" It is said that every man has his price, but in my case nothing ever came close to tempting me to jeopardize my devotion to my military career. I would not jeopardize my retirement and pension for anything they had to offer. I did attend social functions where I was wined and dined by Korean officials as a part of my duty. Most of the surplus supplies were given to orphanages, many of which were created for the sole purpose of receiving the cement and other surplus materials.

It was my duty to investigate these agencies for legitimacy. During my investigations I became aware of the terrible plight that existed for the orphans created by the war. I allowed myself to become personally involved with two of the very legitimate agencies, one was a hospital orphanage and the other a hospital being constructed on orders from General Ridgway, who had been the Supreme Commander in Korea.

The orphanage called the Pusan Children's Charity Hospital was run by an American civilian nurse and much of its sponsorship came from donations from the Masonic Order. I not only approved their supply needs but recruited other officers to accompany me in using them. Four or five officers would spend off-duty time pouring concrete, putting up tin and helping to wash and feed the poor kids.

This orphanage averaged three new babies left on the doorstep a week, some naked or wrapped in rags. In every war zone I was ever in, there was a great amount of prostitution which created fraternization that resulted in illegitimate births. Most of these babies came from liaisons between Korean prostitutes and American servicemen, so they were destined to live in a "never, never" world not accepted by Koreans and abandoned by Americans. Until I saw the miserable conditions under which these poor urchins had to survive, I thought my childhood was rough. They made me realize the value of being an American. The poorest people in the U.S. have life so much better than the people that must survive in a war zone. This seems unbelievable until you have been a witness to both extremes. I learned to not question the morals of these people until I had walked in their paths; I would have liked to have said shoes but most of them had no shoes. I have seen little kids walking in the snow barefoot or with their feet wrapped in rags, scavenging out of garbage cans and dumps for something to eat. The hardest soldier could not help being affected by the plight of these people.

The other hospital which I became involved with was run by the Maryknoll Order of Nuns. General Ridgway, the Supreme Allied Commander in Korea, had issued orders to have a hospital built as a memorial to the troops that had fought in Korea. It was supposed to have been built within a year of his departure. When I arrived, the General had been gone more than three years and the hospital was not half finished. This project had become a comedy of errors which I inherited. The Mother Superior was a very well-educated lady who recorded every promise made by high-ranking officers and dignitaries when they visited the site, but promises was all she got. Another one of the nuns who was an engineering graduate of MIT was actually in charge of the construction and we began working together. She would tell me what was needed and I would try to obtain it for her.

When I started requesting parts for a boiler plant and surgical equipment, I encountered much resistance from the Quartermaster Corps which was staffed mostly by U.S. Department of Army civilian employees. They only took their orders from high-ranking military officers, not a young captain, even though I was filling the position of a Lieutenant Colonel. When I was unable to acquire parts for the boiler system, the Mother Superior wrote a letter to General Ridgway citing all the promises and difficulties she had encountered since his departure. This brought immediate action.

The Inspector General from the Department of Army was sent to investigate the reason this project had not been completed. I had to meet with and orient the Inspector General who by chance turned out to be the Red Snapper. My third encounter with the General whom I learned to admire. After I told him of the difficulties we were encountering, he gave me a written order permitting me carte blanche for anything I ordered. The IG also ordered a Lieutenant Colonel engineer officer stationed with another unit at Pusan to assist me with any technical advice I needed. When I was due to rotate from Korea the project was nearly completed.

One of the last projects I was involved in was the distribution of a ship load of clothing, toys, furniture and other miscellaneous articles that had been gathered in the United States for shipment to Korea. This was a humanitarian project sponsored by Mrs. Ridgway who had been in Korea. While she was in Korea she had been an eyewitness to the poverty and destitute conditions that existed. When I departed Korea, I was given an album consisting of pictures and many Korean letters of appreciation.

About midway through my second tour in Korea, Syngman Rhee, the postwar ruler of Korea was overthrown. Riots occurred all over Korea during the coup d'état. When rioters approached the gates to our compound the CO put me in charge of security since I was the only infantry officer assigned. A Lieutenant Colonel Provost Marshal was in command of the compound but he had never been in combat and his duties involved only Military Police duties. The assigned company of Military Police was placed in my command and we formed a defensive position inside the compound where we could maneuver to any breakthrough point and contain it. Thankfully, the rioting which lasted only three days never amounted to anything more than a few rocks and bottles being thrown at or over the fence around our compound. The worst part of the whole affair in Pusan was the restriction of all Americans to the post and missing many of the Korean employees who failed to show up for work.

While doing my work as the AFAK officer, I was provided an Army sedan with a Korean Augmentation to the United States Army driver. We referred to these Korean soldiers as KATUSAs. All of them were named Kim, although I think one time I heard one called Lee. Anyway, if I took the sedan off post at night I had to get a second KATUSA to run in front of the sedan to keep people out of the way. It had become a familiar practice for elderly Koreans to throw themselves in front of U.S. vehicles hoping they would be killed and the U.S.

government would pay their families restitution amounting to about $45. A small fortune to a Korean peasant at that time.

Another custom I observed in Korea was when a Korean had died, the survivors held a big party where they sang, danced, ate and drank. They then hired mourners to weep and wail as the body was taken to the grave which was usually on a mountainside. The higher stature a person had in life, the higher up on the mountain they were buried. This process was referred to by the Koreans as "Going to the Mountain." One year after a person's demise, his immediate family of survivors would congregate at the grave to mourn the passing. This procedure makes a lot of sense if you analyze it, at death much of the sorrow is dissipated by partying and a year later some of the hurt has passed.

During this period of time, 1958, a practice had developed in Korea where certain Department of Army civilians called DACs, and some enlisted men would become enamored with Korean women. The DACs and soldiers would provide the Korean women with food, clothing, and post exchange items. On their nights off they would spend their time with the Korean women in their huts, called hooches. The DACs had every night off so many of them lived with the surrogate wives all the time. Most of the civilian men were elderly without any family in the States.

These young Korean women administered to their every whim by waiting on them from head to foot. The hooches were small huts with built-up mud floors through which ran four-inch concrete sewer pipes. Charcoal was placed in the pipe openings and lit. This heated the mud floor where the occupants slept under heavy mats. This practice became known as sleeping on a hot floor. I was told it was good for a bad back, but my back at that time never bothered me so I never tried it. Many of the DACs stayed in Korea and married or continued to live with their concubines. It was different with the GIs, when their tour of duty (15 months) was over they would sell or trade their hooch, PX supplies, and girlfriend to their replacements.

As I said earlier, this was considered a hardship tour by the Army, but really the hardship part of it was missing Lois and Skeet and being a witness to the misery endured by the Korean people.

CHAPTER 27

Professor of Military Science

ON RETURNING to the United States, I was assigned as the PMS&T (Professor of Military Science and Tactics) at the Martin branch of the University of Tennessee, where I spent a rather uneventful two years teaching military history to ROTC (Reserve Officer Training Corp) cadets. My boss, a Bird Colonel, who was stationed on the Knoxville campus would come to Martin to inspect my unit every three months, at which time he would sit in on the classes being taught by the different officers.

Once he observed my history class when I was teaching about the Civil War. I asked him to dinner at my home that evening. When the dinner was over and we had a few drinks, Lois asked the Colonel what he thought of my class. He replied, "It was a very good class, but it is the first time I have ever been convinced that the South won the war." This coming from the Colonel, a New York Yankee, came to me as a fine compliment whether he meant it or not. I might interject a saying I have heard: "Save your Confederate money the South will rise again."

The two years at Martin passed rather uneventful. I taught classes three days a week, conducted drill one day, taught small bore marksmanship one day, and spent the rest of the time making tests and grading papers.

During the first summer break I attended ROTC summer camp at Fort Benning for six weeks. The second summer I was exempted and allowed to remain at home in Martin. The entire time I was stationed at Martin, Lois and I were encouraged to participate in community and civic affairs, which we did and enjoyed. Lois, Skeet, and I made many good civilian friends in the counties surrounding the university. Lois and I were regularly asked to be chaperons at sorority and fraternity parties. Many of my cadet officers would regularly visit our home where they would tell Lois their problems and seek her advice. She mothered them like The Bull's wife had mothered us in Japan.

One night we had five cadet officers and their dates out to dinner. The girls they brought were their company sponsors. This was before the government allowed girls in ROTC. One young lady was beautiful and very tall. She was the date of one of the cadet company commanders, who was also a star on the football team. I offered this young lady another helping of food which she declined, whereupon I remarked, "Ginger, (her nickname) you eat like a bird." Then up spoke her date with the remark, "Yeah, like a vulture." Later this couple graduated and married, he became an outstanding high school football coach and she a teacher.

Lois did volunteer work in the Library and Music Department where the cadets would come to pump her about the tests and their grades, because they knew she helped me grade their test papers. I don't know if she ever helped them, but then I never really cared. I often put the test questions on the blackboard the day before a test. The cadets could copy them and look up the answers, since the test questions were the points I wanted to emphasize.

The normal tour for company grade officers (captains or lieutenants) was two years. When my two years was nearly up the Dean of the University wrote a letter to my career branch asking that my tour be extended by a year. When I learned of this letter I sent a letter through my CO to have that request denied and return me to regular Army duty. Lois and I both missed being associated with military people in a military environment. This may sound crazy to some people, but it was the life we enjoyed.

My son had joined the Boy Scouts while at Martin and had progressed to the point of making Eagle Scout at a scout jamboree camp he was attending. I had purchased a load of watermelons and was preparing to take them to the scout camp when Lois called me to answer the telephone.

When I answered the phone, the person on the other end of the line said, "I am Colonel Black calling from the Pentagon. Do you have a paper and pen handy?" I thought this was some kind of joke some of my cadets were playing on me, so I made some bright remark about what kind of game he was playing. Colonel Black then convinced me in no uncertain terms that this indeed was the Pentagon calling. Colonel Black told me they had received my request to return to regular duty and I had been selected for a classified three-year assignment. To start this assignment I was to report to Fort Bragg within two weeks, where I would attend the Special Forces (Green Berets) School. On the completion of that course, I would remain and attend the Special Forces Staff Officer Course. When I finished Staff Officer School I should proceed to the University of

Omaha to attend a semester of prescribed college courses. After that I was to attend language school in California, from there I was to receive orders assigning me to a Green Beret unit on Okinawa.

This was a classified verbal assignment and I would be sent written confirmation immediately. When Colonel Black finished speaking, he asked me what I thought of this assignment and I replied, "What am I supposed to think? You caught me by complete surprise." He told me I had been selected because I was airborne qualified and my experience in Italy. He also told me I would be promoted to Major as they were attempting to upgrade the standards of field grade officers in the Special Forces. With that, he very calmly said "Goodbye" and hung up. Even after he hung up I was still dubious about the call. I escorted the load of watermelons to the Boy Scout camp where I spent the night without telling anyone of the call as I still thought it might be a prank.

Two days later, the orders confirming the call arrived. Only then did I believe it was not a hoax. During the rest of my career, I never received another direct call from the Pentagon.

Lois was pregnant with our second child at the time and since I was going on a temporary duty assignment (TDY) to Fort Bragg, I left her with her mother in Oak Ridge, Tennessee, where she would be near a good hospital.

CHAPTER 28

Green Berets

ARRIVING AT Ft. Bragg, I started the most intensive, rigorous training program ever devised. The primary mission of the Special Forces was to infiltrate an enemy's rear area, contact subversive guerrilla under-ground cells, then train, equip and coordinate the activities of those forces. In effect, establishing a second front that could attack from the rear. The Special Forces to my knowledge had rarely or never been used in what at that time was their primary mission. Due to the fact that the U.S. had not engaged in a linear type warfare since Korea where these tactics would be most effective.

At the time I joined the Special Forces, all members were volunteers. I guess that's why the Colonel from the Pentagon asked me what I thought of my assignment and he accepted my reply to mean I volunteered. I later understood why he said they were attempting to raise the standards of field grade officers in Special Forces. The military was undergoing another RIF (Reduction in Forces) and many officers were volunteering for the Green Berets. They were hoping to overcome some past difficulties and raise their efficiency ratings so they would be retained on active duty in order to complete 20 years for retirement.

I must state here that the officers in the Special Forces class I attended were of the highest caliber, and I would have been proud to serve with them anywhere, anytime. One major called "Shag" had been captured by the Germans in World War II and had escaped. He contacted some French partisans who hid him in a manure bin, and every time he got near a cow pile he broke out with hives. He later became a Major General's aide, and while serving as an aide he lost his wife to cancer. He took to the bottle heavily and the General recommended he volunteer for the Green Berets, so that he could complete his time for retirement. A more courageous person I have never met, yet he had a serious problem he needed to overcome in order to stay in the military.

All U.S. volunteers for Special Forces were required to be Airborne qualified, my reason for stating U.S. volunteers was in order to introduce the fact that we had many Allied Officers in the class who were not required to be airborne. The officers who were not airborne qualified had to rappel by rope from a helicopter.

One exceptional person who underwent Special Forces training, and I hope he will forgive me for using his name, was Mr. Robin Moore, the author of the book *The Green Berets*. Mr. Moore had gotten permission from the government to attend jump school at Fort Benning as a civilian, which he did and then was allowed to attend the Special Forces class which I was attending. To look at Mr. Moore, one might have thought my recommendations for airborne training had been too lenient, but I never saw or heard of Mr. Moore receiving any special privileges in the Green Beret class. His purpose for attending

Special Forces training was to gather information in order to write his book. Mr. Moore attended every class, completed every hike, went on every maneuver, ate snake, set off explosions, fired every weapon, rappelled down the high wall, rappelled from a helicopter and made every parachute jump just like all other trainees did. To many of us younger officers, he was kind of an inspiration. If a middle-aged civilian could take this tough training, we sure as hell could.

The training was very physical. I vividly remember the forced marches where we would go 10 miles in 60 minutes in full gear, weapon, and a 60-pound rucksack on our back. I have short legs, and invariably, I was in the middle or near the rear of the march column that was led by a long-legged training officer. This caused me to over stride or double time, which resulted in raising multiple blisters on the bottom of my heels and these blisters did not heal between marches. One blister would come up on top of the one that had already formed, until finally they got into the red meat and bled real blood not just water. I never failed to make a march because I packed my heels in cotton and inserted a foam rubber pad in my boots. I wish I had asked Mr. Moore if he had experienced the same problem as he and I were about the same stature.

I have discussed some of the officers I met in Special Forces, now I would like to say a few things about the non-commissioned officers I encountered. All the NCOs at Fort Bragg assigned to the school were already Special Forces qualified and hand picked to be instructors. They were truly the elite of the military service. In order to be assigned to Special Forces all officers and men had to meet certain specified qualifications. They were required to be airborne, fluent in a foreign language, be adept in two or more of the following military skills: weaponry, demolition, first aid, communications, escape and invasion, camouflage, transportation, supply and survival tactics. Each Green Beret was cross trained to be highly skilled in at least two of the requirements, in addition to being the most physically qualified men in the service. All the NCOs I encountered at Fort Bragg met all these requirements and more.

During the training, officers were introduced to unconventional warfare subjects such as assassination, sabotage, germ warfare, counterintelligence, and how to utilize available resources to accomplish a military mission. In one of the assassination lectures we were introduced to the possibility of taking a frozen icicle and stabbing a sleeping person in the brain through the ear canal. The icicle would coagulate the blood thereby leaving little evidence as to the cause of death. We were given classes and familiarized ourselves in the use of captured enemy weapons. We were shown how to repair, assemble, and fire weapons captured in Vietnam, Korea, Europe, and the Pacific. We were instructed in ways to utilize readily available resources to make explosives and booby traps. We witnessed a demonstration of the force created when detonating a mixture of a common fertilizer (sodium nitrate) and diesel fuel. The destruction and cratering effect of this explosion was very impressive.

The training consisted of survival procedures that would permit a person to live off the land. We were required to taste grubs, weeds, shellfish, and snake, which led conventional soldiers to labeling the Green Berets as a bunch of "snake eaters." I ate some rattlesnake grilled over charcoal and was told it tasted like chicken, but it did not taste like any chicken I ever ate. I never created a craving for any of the items I tasted, but I could have eaten them (as I did C rations) if it became a matter of life or death.

A problem dealing with counter intelligence was presented to us. Before a Special Forces unit was to be inserted into enemy territory they were to be confined in a top-secret compound, isolated from all outside forces. There they would finalize their special training, study and analyze their mission, then make final preparations. They would have any specialists needed including secret agents assigned to accompany the unit on the mission. One of the questions posed was what would your actions be if you were in route to perform your mission and discovered one of the secret agents assigned was in fact a double agent that could jeopardize the accomplishment of the mission. There was no school answer given to this question, but comments ran the gamut from returning him under guard, to killing him by wrapping him in chains where he could be dropped in the water to sink, to just plain kicking him out of the plane without a parachute. We were told that once committed to a mission, we should let nothing stand in the way of accomplishing the objective.

Had the military been able to use the Special Forces within the concept for which it was originally formed it could have resulted in a tremendous savings in U.S. manpower. Special Forces were structured under a team concept. The basic team called an A Team consisted of two officers and twelve non-commissioned officers. The officers were majors or captains, usually infantry trained and the senior of the two was in complete charge. The team could then be augmented with specialists from other branches of service as needed for the mission. All, regardless of rank, took their orders from the A Team commander. An A Team with needed augmentation was designed to infiltrate an enemy stronghold and contact any subversive underground forces in the area. They then could advise, help organize, train, recruit and equip an underground force of up to 1,000 men and women. The A Team would maintain communication with the theater commander for the purpose of coordinating tactics and procuring needed equipment for the forces they were advising. When two or more A teams were committed to an area, a B team would be formed to coordinate all activities. Two or more B teams in an area would form a C team, which would be the contact unit with the theater commander. In theory, a 14 to 20-man U.S. team could commit and utilize a foreign person force of a thousand in an enemy's most vulnerable areas thus conserving U.S. manpower.

The final training exercise was to be a night jump into a mountainous national park area of North Carolina. In preparation for this jump, all airborne qualified officers made a familiarization jump. The non-jumpers rappelled from helicopters which was far more dangerous than parachuting. Two or three men were seriously hurt when friction from the nylon ropes caused their aluminum snap links to crystallize and break, letting them fall several feet to the ground. I visited one of the men who had fallen, a Filipino captain, in the hospital. He suffered many broken bones, was released from Special Forces training and sent home to the Philippines.

I had been promoted to Major shortly after arriving at Fort Bragg, so Lois decided that before our baby was born she would fly to Bragg for a little celebration. The school gave me a three-day weekend off and I had arranged for a visitors' cottage on the post.

Lois was to arrive at the Fayetteville Airport at about seven on a Friday evening. A storm and fog delayed her flight from six in the evening until nine. At 9 o'clock I was told the plane passed over the Fayetteville Airport, but could not land as the airport had no air ground control system. She was to land at another town about 60 miles away, then bussed back to the Fayetteville Airport. I wore my shoes out walking the floor waiting. She didn't arrive until after 2 a.m. Saturday. She was seven months pregnant at the time and not airborne qualified, but she told me if she had been given a parachute when the plane passed over the airport she would have jumped. Lois stayed at Bragg until Tuesday so that she could watch the familiarization jump on Monday. Then she flew back to Knoxville where she arrived safely. I was not to see her again until completion of the Special Forces School and the Special Forces Staff Officer Course.

The final phase of Special Forces training consisted of a week-long exercise in a civilian area where we were to survive, maneuver, meet civilian people, arrive at and pretend to blow up designated objectives. All this time being pursued by a battalion from a regular Airborne Division stationed at Bragg. This was a field training exercise for them also. Their objective was to capture us while our objective was to elude them.

The class was divided into 20-men teams and one officer was appointed team leader. Another was made second in command and if they or any team members were captured two others would be appointed to take command. If a member was captured he would be returned to the team as a replacement. If the team failed to reach their objectives or any members were captured it counted against the performance rating of every man on the team. I was appointed team commander and Shag was second in command of one team. The exercise started as a night jump into a national forest. The objective was a cleared 50-acre field area surrounded by the forest.

A special Air Force unit had just been formed for the sole purpose of flying Special Forces on insertion type missions. The Lieutenant that flew my team that night must not have received much instruction on Special Forces operations, because he missed the drop zone and hit the green light when the winds were above the peacetime limits for training jumps.

Shag led the jump stick and I was the last man, or stick pusher. When Shag went out the door I heard him yell, "Prepare for trees" but we all jumped. The wind was so high the men could not control their chutes and were scattered all over the area, only a few had hit the drop zone. When I landed my chute caught on the tops of two trees where I was suspended about five feet from the ground. I was able to pop my harness release and fall to the ground. I then set off a signal flare to show the other team members where to congregate. We had several foreign officers on the team and as they were making their way to my position, I never heard so much swearing in foreign languages in all my life. What could have been a major disaster didn't turn out too bad because no one was seriously injured. I still have visions of what might have occurred. This was a remote and unfamiliar area where we had little knowledge of the obstacles present. I was especially concerned about high tension power lines. If a jumper came down astride high powered electric lines he would light up like a Roman candle, and this no doubt would ruin his whole night.

When the team had gathered we set out and reached our first objective where we spent what was left of the night. Each man was required to purchase and jump with what food he needed for a week as no rations would be brought to us.

The civilians in the area had seen several of these maneuvers and they enjoyed playing the game, but we never knew whether they would help us or report us to the airborne unit that was chasing us. I ventured up to a farmhouse where I met a middle-aged man going to his barn and told him I was a Special Forces trainee on maneuvers. He started talking to me and asking questions. When I answered him in his native tongue he called his wife and introduced her. Remember, Special Forces were required to be fluent in a foreign language, and to some people (Yankees for instance) a Tennessee accent (southern drawl) is looked on as a foreign language. We were on the border of Tennessee in the North Carolina mountains, so they had no trouble understanding me. These people allowed the team to sleep in their barn during a rainstorm. While sleeping in the barn Shag broke out with hives, the only casualty we had and it was minor but uncomfortable. The lady left me a quart of milk and an apple pie on their porch. The team reached all their objectives without having a man captured.

I almost forgot to tell about our snake. As team leader I was given an anaconda snake to carry in my pack. An anaconda is a constrictor snake and not poisonous. This snake was supposed to be eaten by the team or kept as a pet. Either way, it was not to be turned loose as they were not native to North America. After the confusion of the jump I decided my team did not need snake to eat and I damned sure did not want a live snake of any kind for a pet. When we got on the ground I cut that old snake into 25 pieces. When the team gathered I told them to take all they wanted, but none was taken so we buried it.

At our last objective we were met by trucks and driven back to Fort Bragg where we were given the next day off to clean and rest up. That evening the team met at a little Special Forces Officers Club bar. We were drinking and celebrating when the pilot that had flown us for the airborne jump walked in and Shag, who was feeling no pain, started telling him what a sorry pilot he was. When the Air Force Lieutenant started telling Shag it wasn't his fault, Shag decked him and we all threw him bodily out of the club. I figured we would all be in big trouble over this, but when it was reported to the Special Forces General called the "Big Spook," he made that pilot fly the same route every night for a week. That was all we ever heard about the fight in the bar.

Upon graduation each of us that had survived the course were given the coveted "Green Beret."

Most of the class had received orders and shipped out in a day or two, but I stayed at Bragg to attend the Special Forces Staff Officers Course which lasted two weeks. This was mostly an orientation course to familiarize non-special forces officers in the use of special forces teams. We were given demonstrations and told of the capabilities and limitations of special operations. One demonstration was put on by a newly formed HALO team. HALO stands for "high altitude low opening." This team used free fall tactics and jumped from great heights using timing devices and oxygen masks. This unit was formed to experiment on ways to insert Special Forces into a hostile environment without being detected by the sound of the aircraft. I'm sure glad I was never used for an operation of this kind. Jumping with a static line is not too scary, but I don't have much confidence in pressure activated timing devices. We used to call paratroopers that closed their eyes when they jumped "night jumpers" and in an oxygen mask I would had been a night jumper. This might have ruined my image as well as spoiling my day.

I was a young, new Major during the Staff Officer phase, thrown together with many senior officers from all branches of the service. I detected some animosity from the regular Army officers regarding the need for Special Forces. I was cautioned by some senior officers that it was great that I had been selected for a tour with the Green Berets, but I should not consider making it a career. Serve one tour and return to regular infantry duty was the advice I was given.

All officers were flown to a coastal Marine base in South Carolina where we witnessed a frog man demonstration and boarded a submarine for a short underwater trip which I did not like. Again, I was glad the Navy rejected me. The Special

Forces had a cross training program with the Navy SEALS, where they were taught to parachute and we were taught to scuba dive. For my part I would rather jump many times to making one dive but to each his own, or what they might say today, "whatever makes your boat float."

CHAPTER 29

Omaha

FINALLY, THE Ft. Bragg phase of my assignment ended and I was given a 30-day delay enroute to Omaha, Nebraska. For those that are inexperienced in the ways of the service, I was charged for leave time even though the course at Omaha was not scheduled to start for 40 days. The delay in route was very convenient to the government, but it was also great timing for me. I arrived in Tennessee two days before my daughter was born. When the doctor came out of the delivery room he said, "You have that little red-headed girl you always wanted." When I saw her, she was all red and very tiny but I was in love with her. Lois had a difficult time with the birth of our son but she came through the birth of Tess with much less difficulty. I served all around the world in many countries, but both my children were born in the great Volunteer State of Tennessee.

There is 14 years difference in the age of our two children. When Lois and I were quizzed about the difference, I would tell everyone I had a big "headache." Lois would say, "I was away so long I forgot what caused babies." At the time of writing this book, Lois and I have been married 47 years and I cannot conceive of us ever being separated except by death. A few words of advice for any young couple desiring to make the military service a career. Military service is not just a job but a way of life which all members of a family must accept. There will be long periods of separation, a lot of uncertainty, constant moves, and much suspicion about infidelity. In the case of Lois and I, the uncertainty of success in the Army was no different than what it would have been as a civilian. Each move we made was considered a new adventure and a chance to meet new acquaintances, or be reunited with old friends. The fidelity is built on trust and the long separations resulted in a new honeymoon each time, or as the saying goes, "You can get so far behind in sex, and get caught up so quick."

While on leave to events that occurred, one good and one bad. The good thing was my daughter being born. The bad thing was the assassination of President Kennedy. I had just come home from a fishing trip with a friend when Lois met me as I got out of the car and gave me the news, "The President has been shot." I just could not believe a thing like that could happen in our country. President Kennedy was a very strong advocate of the Green Berets, as well as being my Commander in Chief. He was the second president to die in office during my military service, the first being President Roosevelt. Even though I had never personally seen either of them, their deaths left me with a great sense of sorrow. It was as if I had lost a close friend.

All too soon, my 30 days leave was over and Lois, Skeet, Tess and I were off to Omaha where I enrolled in school. I was required to take a humanities course and a public speaking course. I suppose the Army wanted me to gain some culture. I also took a course in television and one in driver's education. The only humanities course open was "Arts and Crafts for Teachers."

In the beginning as most people would, I figured this would be a "crypt" course, but how wrong I was. I got off to a bad start with the instructor who was a very pretty lady in her early thirties. The first day of class the students were all sitting around work tables while she was demonstrating how little students could use scissors to make art works. I started looking around at my classmates, many of whom were military men my age or older, and I guess I was grinning instead of paying attention. I was thinking to myself how funny I must look: Me, a grown man, a Major who had just finished the most rugged training course the military could devise. Now I was sitting around cutting out paper dolls. All at once the instructor stopped talking and looking directly at me she said, "Mr. Walker, if you think this class is funny you can leave now." I assured her that I did not think her class was funny and that I was thinking about something else, to which she sternly told me to pay attention. I learned later that she had been involved in a love affair with a Navy officer who had jilted her. Leave it to the Navy to foul things up.

This class was meant for future elementary school teachers and we were required to use as many different media forms as possible. We painted with tempura, carved soap, formed paper mâché animals, built mobiles, and did a collage, plus many other art forms. Each week we were required to turn in a sample of the medium we were using that week. The instructor would then select some of the best examples and place them on display in a window case. My examples made the window each week and my classmates started calling me "teacher's pet." How juvenile can one get? We were required to make a scrapbook containing all of our examples, then write a short essay on each explaining what materials were used and how we used them. For the final exam we had to write a 30-page monograph on the "Theory of Art." I was never sure I understood the Theory of Art, but I passed the course. A bad start sometimes has a good ending and the main thing I learned in her class was that if you wanted to teach a mule something you had to first get his attention. This she did to me the first day of class.

Another course that I figured would be very difficult turned out to be very easy for me, that was Public Speaking. I had given a lot of instruction in the Army, been an Assistant Director of Instruction at the Infantry School and taught Military History at the University of Tennessee, Martin. So I had a background that enabled me to breeze through the course. I was the only military person in the class. All the other students including the instructor were much younger than I. Each student had to give a demonstration lecture about a subject of their choice. I chose to demonstrate how to cut up a chicken for frying, Southern style. I got the attention of the class when I poured myself a martini, which I said one would need before attempting to cut up a chicken properly. After my demonstration I discovered none of the young students at ever cut up a frying chicken. They thought they always came that way from the supermarket. And these were young college students who came mostly from the heart of America's farmland.

Not very much exciting happened while I was in Omaha. My daughter was growing and it gave me more time to spend with my son, Skeet. I met some people who took us pheasant hunting which Skeet and I really enjoyed. Lois and I met two Air Force nurses living and our apartment building who enjoyed taking care of Tess, and that allowed Lois and I to have some time to go out together. Just before the semester ended I received a change to my orders. Instead of going to the language school in California, I was to report immediately upon completion of classes to the military district of Washington, DC where I would receive further orders. I was not supposed to remain in Omaha for the graduation exercises and I was given only three days to arrive in Washington. I learned of the need for urgency when I reported in at the military district office in Washington. However, since this was a classified assignment, I don't feel I can discuss it in this book.

CHAPTER 30

Washington, DC

ARRIVING IN Washington, I was told I had been selected for a classified assignment that would be a non-efficiency reporting period. I was given a year advanced pay vouchers which I could redeem at any military finance office for drawing my pay. I was then told to wear civilian clothes and report to a Navy officer who would give me further instruction. I was told not to report back to the military district offices.

Then I met my Navy contact, he told me to report to an office in a building in downtown Washington, DC. There I would be met by a Sergeant and a Burmese lady who would tutor me in the Burmese language. I was not to contact the Navy unless they contacted me first, in other words he said, "You are on your own." The Sergeant and I met with the Burmese lady almost every day for nearly a year where we were able to learn enough of the Burmese language to hold a conversation. We also learned about many Burmese customs and the political structure of Burma at the time. Burma had been an old English colony that had gained independence. At that time, the Burmese government was controlled by two factions, one pro-Chinese and the other leaning towards the U.S. "U" Thant was the head of the United Nations at this time which left our government in a delicate position when making any policies concerning Burma. Burma is made up of many ethnic cultures or tribes. I was told there were over 100 languages or dialects spoken in Burma and less than one tenth of the Burmans would be able to converse with me in the language I was being taught.

The Burmese lady who was my tutor was also the wife of the Comptroller for the Burmese Embassy. A part of my assignment was to cultivate a friendly relationship with the Burmese Embassy people. This part of the task was very enjoyable and easy. Lois and I visited their home where we became very good friends with my tutor's husband, her brother, his wife, their children, and an aunt who was an American citizen who worked at the Smithsonian Institute. I met the Ambassador and all the staff and was very impressed with the Burmese military attaché. The tutor and Lois became very good friends. Lois would invite the whole Burmese family to our home where the ladies would exchange recipes, then teach each other to prepare Burmese or American food which we all ate. U Thant's niece became a regular visitor to our home when she had a break in her college classes.

On the anniversary of Burma's independence an all-day celebration and party was held at their Embassy in Washington, DC. This was a very lavish affair to which Lois and I received a formal invitation. Because I spoke the language and had met so many of the people, I received preferential treatment. I complimented the Burmese ladies on how nice things were arranged and how delicious the food was. This was a practice not often given by Burmese men. The little ladies went out of their way to serve Lois and I food, even to the point of pushing a high-ranking U.S. officer out of their way. Someone asked the Burmese attaché how long "U" Walker (U in Burmese is a title bestowed on a respected person, hence U Thant) had been in Burma. He replied that I must have been there for some time (even he did not know that I had never been there) in order to speak the language so well. The person who asked the question agreed with the attaché that I must have been there a long time, but he also said Mrs. Walker must never have visited Burma as she did not speak Burmese very well.

The rest of this assignment is classified so I will leave this subject. When I was leaving Washington I had the privilege of sponsoring the Comptroller, his wife, their children, his brother, and sister-in-law to citizenship in the United States. Later the brother and sister-in-law had a baby girl and they named her after our daughter Teresa, whom I call Tess.

The only problem during my time in Washington was gaining weight. After the rigorous training at Fort Bragg, my time in DC was almost sedentary. I would get up before daylight and run a mile or more every morning. This was before jogging became popular and I was afraid people would think I was crazy running around Washington, DC, but in retrospect I don't think it would have mattered as crazy seemed to be the norm in Washington even then.

CHAPTER 31

Germany

MY ASSIGNMENT over, I was told to report to the Infantry Career Management office in the Pentagon where I was allowed to choose an assignment of my choice. As Lois had never been to Europe, I elected to go to Germany. The only thing they had to offer at that time was an Inspector General slot in Munich, which I took. I had a two-week orientation school at the Pentagon where I was voted by the class to be the most successful Inspector General in the class, but as you will see this accolade wasn't very accurate.

Lois, the kids and I were scheduled to sail to Germany on a luxury liner, but Lois because of her seasick experience coming home from Japan told me I could take the kids on the ship and she would fly over to meet us. Tess was under two years of age at the time and Skeet was under 16. I could not see myself taking care of these two youngsters by myself, so I went to Career Management and asked them to change my orders to allow us to go by air, which they did. That's the only time in my entire career that I ever asked for exceptions to an order. Many times I questioned orders, but once confirmed I always complied to the best of my ability regardless of the consequences.

When we arrived in Munich I encountered the worst situation I was ever in. I was assigned to the Munich Area Command, an administration and housekeeping organization filled with some of the worst senior officers I ever met. This command had been in Munich, then discontinued and moved to Wiesbaden, reactivated and moved back to Munich. I arrived during the confused reorganization from Wiesbaden back to Munich. Because of my previous special assignment, I was given concurrent travel with my family and was to be given immediate housing in Munich.

The IG section consisted of seven majors and an old "fuddy-duddy" Quartermaster Lieutenant Colonel. The command had a general officer whom I never met, as he stayed in Wiesbaden during my short stay in the command. I was assigned to the inspection section with four other majors, one of whom was designated chief of the section. I will call him "Beebo." Beebo outranked me by two months so he was my rating officer. He had been in the command two years where he had housing for his family in Munich and was forced to move them to housing in Wiesbaden. Now that he was being returned to Munich he was again awaiting a housing assignment there. He did not take kindly to my orders directing me to receive immediate quarters in Munich.

I was ordered to accompany the inspection team on a three-day trip the day after I arrived in Munich, and my family was put in temporary quarters where they were left to fend for themselves. Beebo had never been in combat even though he had 16 years of service in the infantry. He had been a spit and polish officer in the "Old Guard," the ceremonial unit in

Washington, DC where he had marched in funeral parades. Beebo still wore steel horseshoe shaped heel plates on his boots which he stomped around in as if he were a storm trooper. He had not completed the educational requirements which at that time the criteria was for company grade officers to have a college degree or to be taking classes towards obtaining one, which he was not doing. Two of the other majors on the team had been classmates of mine in the Infantry Advanced Course. As you can see Beebo and I were not hitting it off well.

When I returned from the inspection trip I asked for a seven day leave so I could get my family settled and visit my brother-in-law. The one I had sworn into the Air Force in Knoxville, who was now a Lieutenant in the Air Force stationed in Wiesbaden. Beebo had to approve my leave request which he told me that he had done, but when I went to pick it up I found it had been disapproved. Whereupon, I went to the Lieutenant Colonel and demanded why it was disapproved. Old Fuddy called Beebo in and asked him why my request was denied. Beebo said he thought I needed more work on writing inspection reports. Well, this led to a real confrontation. I offered to whip Beebo's behind and the old Fud got at all excited and cried we couldn't have this kind of thing going on in the Inspector General's office, as he was charged with inspecting and reporting acts of this nature. This all ended with my taking seven days without their approval and when I returned I was told I was a troublemaker and I should find another home.

There was a Special Forces unit stationed at Bad Tolz near Munich, so I paid them a call. This unit was commanded by a full Colonel and when he looked at my records he told me with my qualifications he not only wanted me, but he was going to insist on my being assigned to his command. I could keep my housing in Munich because they were also allotted housing there. When I returned to the tension-filled IG section, Beebo and Fuddy had decided that if I didn't have to give up my housing, they weren't going to let me go. Instead they sent a special efficiency report to Eighth Army Headquarters giving me a zero rating which eventually came back to haunt them.

About 10 days later I received a call from a Colonel I had known at Benning who was now the Commander of the post in Verdun, France. When I answered the phone, Dan (the Colonel) said, "Art, a friend of ours in the Eighth Army Personnel section called and said you were having a little trouble in Munich. Would you like to come over as my deputy?" Before I could answer he said, "All I can offer you is a General's quarters, a full staff, and to put you in complete charge of the area." Naturally, I could hardly contain my elation, but I was able to ask, "How soon?" To which he replied, "I'll have your orders to you within a week." When Old Fud and Beebo heard I was getting orders to France they hardly spoke to me. The rest of the IG section offered their congratulations and asked me to try to get them reassigned. I went home where Lois and I started packing immediately.

We had purchased a Volkswagen Beetle which we loaded down with luggage, personal items, two kids and a dog which had been given to us by one of the other majors on the inspection team. We had purchased a grandfather clock from an old German estate which I had to tie on the top of the Beetle. We must have looked like gypsies as we drove the Audubon from Munich to the French border. Every time a big German Mercedes passed us it nearly blew us off the road. The Germans had only two speeds on their cars, complete stop or as fast as it would go. Germany was a very clean country with super highways, and you rarely saw any trash thrown around. I wish I could say the same about France, but it would not be true, so I choose not to elaborate on this subject.

CHAPTER 32

France

WHEN WE arrived at the post in Verdun, Dan and his wife met us and had arranged temporary quarters in the guest billets. They had a table set up in the Officers Club dining room with a sign saying, "Reserved for the Deputy Commander." All this protocol was new to Lois and me, so when we went in to dine we were observed with quiet reserved and inquisitive glances. Tess started crying loudly as soon as we were seated, and this seemed to break the tension in the whole room. At first a couple of nurses came over and offered to help and ultimately most of the patrons came and introduced themselves. I did not know at the time that Dan had already told people that I was a mean Green Beret and had given me my second nickname. He called me the "Tiger." If I were a tiger I never had to really bite anyone in Verdun, scratch them a little maybe, but I never bit anyone. The Colonel was a long-time infantry officer with a good record and true to his word he put me in charge of running the post.

Verdun was a big operation. We had over 800 dependent housing units, a hospital, a dependent school, and many supply depots. We were administering to the needs of five tenant units of considerable size who did the supply ordering for all of Europe. My job was to coordinate all the operations of the entire command and I had a complete general staff to assist me. I had a very efficient American civilian secretary, a French protocol officer, and a detachment of WACs (Women's Army Corp).

During the two years at Verdun, I encountered no major problems and only a few minor ones. Most of the minor problems were engendered by military dependents, namely mine. The first thing I did every morning was to look over the Military Police report to see what problems had arisen during the night. One morning the first name on the report was my wife, Lois. She had been given a citation for not stopping at a stop street. Since she was not military there was little I could do except write her a letter of reprimand which I did not do. I did call her to tell her not to let this happen again. I told the Colonel about what had happened which gave him a big laugh, then he called Lois and his wife to meet us at the club for lunch where he chided her about the citation.

Another instance involved my son and the Colonel's daughter. Both had just started high school and they were very good friends. I was very strict with my son as the Colonel was with his daughter.

Before I arrived, my predecessor had talked the Colonel into establishing a teen club for the dependent children. My son did not go there often but the Colonel's daughter did. One night she didn't get home until after 10 o'clock, the time the club was to be closed. The next morning he ordered me to close the club for good, as he thought it was not being properly supervised. I became the "heavy" with the kids and some of them tried to give Skeet a hard time because his old man had torn up their playhouse. My son took it in stride as I was permitting him to go to Italy on his own to attend the International Boy Scout Jamboree. While there he was tapped out for the "Order of the Arrow" which is one of the highest awards bestowed on a Boy Scout.

The time spent in Verdun was very gratifying. The area was steeped in history of World War I and World War II. The family and I had time to tour the area. We visited the ruins of a church at *Douaumont Ossary* where the bones of World War I soldiers were left piled in the basement. Just outside was the *Trench of the Bayonets* where dirt had covered dead soldiers while they were standing which left their bayonets still protruding from the trench. This place had been the left by the French Government as a World War I Memorial. We visited the fortifications of the *Maginot Line*, the supposedly impenetrable fortification of World War II.

Barley Douke, made famous in a World War I song, was only a few kilometers away and we went there often.

When I had been in Verdun nearly 2 years, General de Gaulle kicked the U.S. Army out of France, so all the installations were moved to Germany. Colonel Dan rotated and left me in command for about a month. As the commander, I was invited by the French Commander in the area to review the French troops and World War I veterans on November 11, "Armistice Day" which is still called Veterans Day in France. I considered this a big honor and my picture even appeared in the Paris newspaper.

Lois and I really enjoyed our tour in France where we were able to make many "junking" trips. Junking was a term given to hunting for and buying foreign antiques. We would visit old homes where we would bargain with the residents for their antique possessions, some of which were valuable, although many worthless.

After being in command of Verdun a month, I received orders curtailing my European tour and was to return to Fort Riley, Kansas to help form a newly activated Infantry Division deploying to Vietnam.

CHAPTER 33

Fort Riley

I HAVE arrived at the time in my military career which is the most difficult to write about. The events I have related in World War II occurred over 50 years ago and time has a way of dulling the memory. I came home from WWII to a country that had been united in a common goal. Every American had participated in that war in one way or another. Every American had a strong sense of pride and patriotism which they willingly showed. I felt the pride of all Americans. They would salute the flag. They would stand and uncover when the Star-Spangled Banner was played. They pledged allegiance to the flag. We wore our uniforms with great pride on ceremonial occasions. Surplus military equipment was restricted and not sold everywhere, so it retained its symbolization of patriotism.

My Korean experiences happened 40 or more years ago, and I felt a small rift in the feelings of Americans, after all this was not _our_ war. As a matter of fact, it wasn't even called a war but labeled a "police action." By now having been drafted I had decided on a military career which put me in a different category. Yes, I was still proud to serve my country, but I was also serving myself. It was a way to make a living and support my family and I had chosen it as my profession.

Many recalled veterans found themselves in a like category. Their civilian lives had been disrupted, some for a second time, others who had been drafted for the first time returned to civilian life, but not to the accolades and benefits shown World War II veterans. Korea became known as the Forgotten War with the United States neither claimed victory nor defeat. In these two wars as in all wars, everybody really loses because of the great losses in human lives, the desecration of human bodies, and the wanton waste of resources.

Arriving home from France I took a 30-day leave which I spent building a home in Tennessee for Lois and the kids to live in while I was in Vietnam. Lois and Tess accompanied me to Fort Riley, but we had to leave Skeet in Tennessee with my father-in-law to start school. We knew I would only be at Fort Riley a few months and then Lois would return to Tennessee.

At Riley I was assigned as S4 of the first brigade of the division with the task of organizing and training a supply section of newly enlisted or drafted soldiers that had barely completed basic training. During the training phase the brigade was put in command of three infantry battalions, an attached tank battalion, and an artillery battalion.

Altogether the brigade consisted of approximately 5,000 soldiers who needed to be trained, equipped, and molded into a coordinated fighting machine ready to move to Vietnam and be committed to combat. This had to be accomplished in less than six months.

The officers and non-commissioned officers had been brought together from all over the world and in most cases had never worked together as a unit. Very few of the junior officers had seen combat duty and most of the senior officers had been in desk jobs or service jobs like the one I had just completed.

It was taught in the infantry school that a commander "is responsible for everything his unit does—or fails to do." You can imagine what a heavy burden this places on an officer selected for command. Yet, infantry officers fought for command slots, because if they were successful their place in the military was assured. If they were unsuccessful they were court-martialed or faded into military oblivion. The Colonel selected to command the first brigade was fairly young and had completed all the necessary schooling. He was commissioned from college ROTC, had attended the Advanced Infantry School, the Command and Staff School, then graduated from the Army War College. Maury, (his nickname) had been an instructor at the Infantry School where he had been well indoctrinated in command procedures and tactics, but he was lacking in the field of logistics. Remember the quote on logisticians.

Shortly after my assignment the brigade held a review parade that did not come up to Maury's standards. He gathered all the officers together on the parade field where he proceeded with a mass "ass" chewing. When he finished I asked to speak to him in private, where I bluntly told him that I could not work for him. I had been assigned to do the logistics job for his brigade and when I failed to do my job I would accept any criticism or efficiency rating he wanted to give me, but I would not accept generalized criticism. From that day on Maury left all logistics problems to me and I never received another "ass" chewing. I wish to say here that Maury was one of the better officers I ever worked for. He went on to make Major General and I hope maybe Lieutenant General. Just making General with an ROTC commission at that time was a big accomplishment as most generals were military academy graduates.

Maury was a stern task master which was needed to form a cohesive unit soon to be committed to the process of killing in Vietnam. Most of the soldiers were young draftees or people that had enlisted in the service much as I had in World War II. In many respects I sympathized with them but I knew I could never let it show.

I had a young, just out of high school, soldier assigned as my driver, with whom I was very strict. Once he ran out of the road and nearly threw me out of the jeep. For this I really gave him "hell." Lois heard about this incident and she really laid it on me; she reminded me this boy was nearly as young as our son. She also told me how she had observed the devotion this young boy had for me. He would wait hours for me and do anything I told him. She really laid a guilt trip on me, but I never let it show because I knew what was ahead for us in combat, as I had been there before.

The young soldier's name was Percy (not real) which caused the soldiers in the headquarters company to call him Pussy. The name Pussy may have resulted because of his devotion to his duties and to me. Pussy was with me the entire time I was in the brigade and he would foul up, which wasn't often, I would tell him I was going to transfer him to a rifle squad, but of course I never did. When I left the brigade in Vietnam I was able to get him transferred to a clerk's job in Division Headquarters; this was about the safest place you could find in Vietnam.

Another guilt trip laid on me was created by a first cousin of mine, whom I had not seen since he was a small boy. I was inspecting the Brigade Air Section one day when I noticed the name tag on one of the young soldiers read "Walsh." I recalled my father's only sister's married name was Walsh, so I started asking him where he was from and when he replied, "Indiana," I asked, "are you from Moreland and is your mother's name Mary?" He was taken by surprise, but he answered, "Yes." I told him his mother was my aunt. About 10 days later I received a letter from my aunt asking me to take care of her only son and keep him safe. Something I could not promise. This soldier became a very good helicopter mechanic and was promoted to Sergeant, where he was well liked and respected by all the helicopter pilots. He returned home from Vietnam without being physically wounded, but after his discharge he suffered the same remorse that was prevalent among most Vietnam veterans. It's a terrible mental burden for a young person to bear being criticized by the very people whose rights he was giving his all to protect. I will never understand the stigma attached to the people who served in Vietnam.

About three weeks before the unit shipped out, Lois returned to Tennessee leaving me a bachelor at Fort Riley. Several of the officers whose wives were going to remain at Riley after the Division left would ask me to dinner. I became acquainted with their wives and children which caused me much sorrow later. Death and wounds are a part of the expectations encountered in the Army, especially in the Infantry. When these actually occur, it is easier for a Commander to notify the dependents if he is not acquainted with them.

Maury was designated to command the advanced party for the Division and I was among the hundred or so selected to accompany him. The task of the advanced party was to fly to Vietnam ahead of the Division and once there, establish a secure base camp in preparation to receive the main body of the Division which was coming by ship.

Our Division was part of the big buildup of forces in late 1966 and '67. Our flight took almost 50 hours in a military cargo transport plane with canvas bench seats and box lunches. We carried all our weapons, ammunition, and combat gear on our backs or stowed under the seats. The only time we could stretch our legs was when the planes landed to refuel and even then, we were not allowed out of the sight of the plane.

CHAPTER 34

Vietnam

WHEN WE arrived at Tan San Nhut Air Base, we were met by an armored tank battalion that escorted us 25 kilometers to a base camp under construction. This camp was in a swampy jungle area that was being cleared by an engineer Battalion using huge bulldozers with a curved plow shaped blade called Rome Plows. Many of the advanced party, including myself, were subjected to Agent Orange when the Air Force sprayed the area to defoliate the trees before the Rome plows dozed them over and piled them in long rows. Maury selected one of the fringe areas for the first brigade's camp and put me in charge of getting it ready to receive roughly 5,000 soldiers that were to arrive within two months.

The base camp covered several hundred acres that had to be cleared. A berm (dirt wall) was pushed up completely around it to be used as a defensive parameter which had to be guarded at all times after the entire division arrived. This base camp was a safe haven, but in my opinion a violation of military tactics which I will have more to say about in the closing chapters of my book. Initially, the advanced party had to provide their own security, but we only experienced a few probing attacks by the VC (Vietcong) which were easily repelled with aid from the engineers and the armored battalion.

Base Camp Viet Nam 1966

The monsoon season arrived at about the same time as the main body of the Division. The area turned into one gigantic swamp. The tents which had been set up on the ground had to be raised three feet on wooden platforms. This caused me difficulty in the scrounging enough lumber as every other brigade was doing the same thing. Luckily, a warrant officer who had worked with me in the experimental division at Fort Benning was in charge of building supplies in Saigon. He turned his back while I loaded 15 trucks with lumber and three dismantled Quonset huts (metal buildings). He probably wrote the lost off to theft by the Vietnamese, anyway the First Brigade had the first metal building mess hall in the division. Maury never asked where they came from and I never told him, but there were questions by others in the Division who never got an answer either.

The following pages contain excerpts from printed newspaper clippings taken from a scrapbook which enables me to describe some of the actions taken by the First Brigade. Again, I avoided using unit names or names of people, other than my own, to protect their identity in the event they failed to return from Vietnam. Those that participated in the heroic efforts will be able to recognize their actions and if they wish to be recognized they should contact me.

(Long Than) "The old Reliable 1st Brigade, the most traveled unit in the division, has fought in operations from Vung Tao to Tay Ninh since its first elements touched Vietnamese soil on the last day of 1966."

"The highly mobile battalions of the first and foremost have fought together as separate units under operational control of various higher commands. Participating in one or more of the following operations were the 2nd and 4th Battalions, Infantry, 3rd Squadron Armored Calvary and 1st Battalion Artillery."

"Colby – Infantrymen fanned out from Base Camp in search and destroy maneuvers, made the first significant contact with the enemy and killed fourteen."

"Big Springs - Numerous bunker complexes and food supply depots were seized. First Brigade civic action programs intensified. Five enemy vehicles were captured in the operation conducted in north east Bien Hoa Province. Large caches of rice and ammunition were uncovered."

"Greenleaf - One of the first joint American-Vietnamese operations for 1st Brigade units, the action was marked by search and cordon in which U.S. forces would seal off a village so Vietnamese troops could search out the enemy. The operation took place 10 miles south of -- Base Camp."

"Junction City - Largest ground operation to date of the Vietnam war. The first major encounter of the -- division units and hard corps Vietcong soldiers to 228 enemy dead after a furious six-hour battle five miles north of Lai Khe in Bin Duong Province. Elements of the -- Calvary 'Black Nights' withstood an onslaught of mortars, rockets, automatic weapons, and small arms and severely battered the 273 VC Regiment. The -- Infantry Battalion also made solid contact with the enemy during the operations."

"Port Sea conducted near Vung Tau, Operation Port Sea resulted in 47 enemy killed by Old Reliable. The joint operation --- with Australian forces lasted 12 days."

"Manhattan Junction City III – A month long operation during May was aimed at clearing out enemy portions of Tay Minh Proving."

"Now once again for the first time in months units of all 1st Brigade battalions are fighting together as a Brigade in operation AKRON near the divisions base camp." *(End of Excerpts.)*

You can see from these excerpts why the 1st Brigade became known as the "Fire Bucket Brigade," we were sent where ever a problem with the Vietcong occurred. I have many more clippings detailing the actions in depth, one of which I will include as it concerns one incidental action on my part. I received the "Distinguished Flying Cross" for just doing my job.

"<u>Close-up on Colby</u> – 2 Majors Make 11 Sorties to Resupply Unit. Supplies can mean the difference between life and death in combat. For Company A (unit omitted) during Operation Colby they meant just that. Deployed in an area difficult to reach by land, the unit went 36 hours without receiving additional supplies. To ease the pressure of the supply problem, Major _____ air liaison officer and Major Arthur R. Walker, Brigade S4 (logistics) officer braved enemy fire for six hours. The officers flew 11 sorties in a helicopter to deliver essential supplies, evacuate the wounded and bring back intelligence information." *(End of article.)*

I think the award of the "Distinguished Flying Cross" was a mistake and came as a surprise to me when I was called out to receive it. I may be the only non-flying infantry officer to receive a medal normally given to pilots or aircrew members. I was later told it may have come about because of faulty radio communications, as the pilot was to receive the DFC and I was put in for the DSC (Distinguished Service Cross). I relate this incident to illustrate how a <u>few</u> actions were observed and awards given but <u>many,</u> <u>many</u> more far braver actions went unnoticed or unrecorded.

Now I will reveal one other action for which I received much satisfaction, but I did not want it recognized or recorded. This action concerned a shipment of Maine lobster.

While on an operation near Vung Tao, I was in a food supply depot that had just received a big shipment of frozen Maine lobster, a delicacy of which I am particularly fond. This shipment was intended for the commissary store in Saigon where the big brass and U.S. civilian employees purchased their food. Since Saigon was off limits to my division at that time, I knew my units would be unable to partake of the delicacies. I felt my brigade was more important than anyone else and deserved the very best.

I knew we were about to complete the action and were going to be deployed north to Tay Ninh. I arranged with the people in charge of the depot to have the lobsters diverted to my headquarters in Tay Ninh. I will never reveal the names or ranks of the people in charge, but this little error on their part only cost a few captured enemy weapons. When the Brigade arrived at Tay Ninh three days later the entire Brigade feasted on Maine lobster. I was invited to three different messes where I consumed a lobster in each. To my knowledge, this was the only occasion lobster was ever included on the troops menu. The recognition and satisfaction I got from this quasi illegal action was worth more than any medal.

Captured VC Weapons

Another action for which I received an award involved a Tank Bridging unit one of only three in the U.S. Army, which cost $50 million a copy. This is beside the point, but I think the military was vastly overcharged for these monstrosities. It was an oversized tank with a telescoping bridge attachment which could be deployed across a 30-foot ditch or stream for other tanks to cross on. One of these Tank Bridging units had been assigned to the Brigade's Calvary (tank) Battalion and while attempting to deploy the bridge, the tank hit a mine in the middle of the stream which blew off one tread, thus dismantling it. It was too heavy for their task recovery vehicle to dead drag out of the stream. I was sent to determine whether it should be blown up and abandoned but knowing the cost I was reluctant to destroy it. With the aid of some very good NCOs we devised a scheme to drag the damned thing out of the river where we could replace the tread. We formed a chain by hooking three armored personnel carriers and the tank retriever to the bridging tank after removing the bridge part. With some excellent synchronized driving on the part of the personnel carriers and retriever drivers they were able to drag the bridging tank to dry ground without hitting another mine. We had already removed two other mines from the stream. Once on dry ground the men were able to replace the damaged tread. It had taken most of the day and we had to return to the base camp after dark. This was very scary as we had little protection and a very small force. We all made it back safely, although we heard several small arms bullets hit the tank. I recommended each of the men involved for the commendation medal, which they received. Little compensation for saving the multimillion dollar junk pile. Armored personnel might question my evaluation of the bridging tank, but what do I know as I am only an infantry man. Tanks are handy to have in some situations, but infantry is always needed.

In order to maintain continuity in combat units, it was the policy of the division to reassign half the commanders and key staff officers halfway through their tour and replace them with newcomers starting their tour. By doing this the command would not experience a complete turnover all at once.

Maury had already gone up as an Assistant Division Commander and had been promoted to Brigadier General. The new brigade CO had approved a seven-day rest and rehabilitation leave for me. I was going to meet Lois in Hawaii. It was Division policy to grant these leaves after six months in Vietnam. The military would fly us to Hong Kong, Australia, or Hawaii. I chose Hawaii where Lois could fly at her expense and meet me.

I was to make reservations at a hotel in Hawaii for a Major and his wife who was to take R&R when I returned. This Major had just been appointed Brigade S3. I had met his wife and kids at Fort Riley where I had been invited to their quarters for dinner after Lois had returned to Tennessee.

The day I left to go on R&R I brought the bodies of the Major and a Lieutenant Colonel Battalion Commander to Saigon in body bags. They had been killed by a sniper just an hour before I was to leave. This has been the most difficult part of my story and has been the cause of my not writing it for 26 years. I have been told I should name these two friends as therapy for myself, but I find I cannot do it even now for I fear I might rekindle the sorrow their survivors have endured. I have never visited the Vietnam Memorial Wall where I would be able to see theirs and other names engraved. I have no need to see that wall as I have those names indelibly engraved on a wall in my mind.

Lois had arrived in Hawaii a day before I was to arrive and I knew she was there waiting, so I did not take kindly to nor adhere to the Army or Customs procedures that had been established for personnel arriving from Vietnam. To me they were only delaying actions. When I went through Customs they wanted me to empty my duffel bag so they could search it for contraband or drugs. I told them they could keep the damned thing because I didn't need it, whereupon the Customs agent said, "Go ahead." Then a Sergeant told me I had to attend an hour orientation lecture on how to conduct myself in Hawaii. I pointed to my oak leaves and told him, not to politely, what he could do with his orientation lecture. I left him standing there trying to round up others for the lecture, but it did not seem he was having much success. When I got to the hotel where we were to stay it was nearly midnight, but Lois was up waiting for me and the second thing I did was drop my duffel bag.

Lois had rented a car for our use, but we never used it and we turned it in the third day we were there. I actually rested and rehabbed myself for seven days. Lois and I were staying on the beach and for seven days we walked, wadded, or sunned, even though I never needed to sun after coming out of the heat and jungle. The mean average temperature where I was stationed in Vietnam was 110°. Lois and I wore bathing suits most of the time, which was a big relief from the jungle fatigues which were nearly always wet with perspiration and showed big white salt rings under the arms. Lois had brought some of my civilian clothes for me to wear while in Hawaii which she had to take home when she returned, as we were not allowed anything but combat gear in Vietnam.

We went barefoot so my feet were allowed to dry out after being constantly soaked in the swampy jungle. When the troops returned to base camp after an operation they were allowed to wear sandals to dry and air their feet to prevent what is now called immersion foot, but in World War I and II it was known as trench foot. Even with this precautionary measure, the Brigade suffered many casualties though they were minor in nature. If the readers were to soak their hands in dish water all day and see how wrinkled they became this would give them an idea of what immersion foot was like. Do this three or four days in a row and you would find your hands really sore.

After the sad experience I had coming out of Vietnam for R&R in Hawaii, I debated with myself about whether I would return. I even mentioned it to Lois who reacted as though she thought I was kidding. She did tell me when I went back not to be a hero, which is something I don't think anyone ever plans.

Three days before I was to return to Vietnam, a supply warrant officer who worked in my section arrived in Hawaii on R&R. He informed me the Division had selected me to be an advisor to a newly formed volunteer force that was formed for duty in Vietnam. The seven days of R&R went by the quickest out of any time in my life and saying goodbye to Lois was almost more than I could do. I managed not to show all my emotions but I'm sure she felt them. She is a strong person and was able to put forth a big front for my benefit. The long flight back gave me time to think. What if I had failed to return to Vietnam? I would have been a deserter, something for which I could never have forgiven myself. I would not have been able to live with or ever face my family and friends again. It would also have put me in the same class of the people who left the U.S. to avoid doing their duty.

Now that I am about to leave the Brigade I wish to convey my thoughts about the very high caliber of men I served with. In 20 years of service I had never seen more intelligent, responsive soldiers. I have heard stories of drug use, fragging officers, misconduct and other major soldier infractions, but I never encountered any of these actions during my stay in the First Brigade. Sure, there were some minor infractions of rules, but when it's all said and done I was extremely proud to have served with these fine soldiers.

When I arrived back at the Vietnam base, I was told by General Maury that the Division had been selected to sponsor a regimen of Thai volunteers who were coming to Vietnam. These volunteers were being trained by a Special Forces team on the Thailand border where it joined Cambodia, Laos, and Burma. I had been chosen for this assignment because of my Special Forces background and the fact that I could speak Burmese. The language was of no use because Burma was neutral, and all the other countries had their own languages.

CHAPTER 35

Thailand

THE ADVISORY team consisted of five officers and fifteen NCOs, each selected for special skills and abilities. For instance, the Lieutenant Colonel in charge of the team had been a Battalion Commander and was to advise on tactics. I was second in charge and I was to be the logistics advisor, other officers on the team had been company or platoon commanders. The NCOs advised their counterparts in their particular skills. The team was sent to Thailand for 30 days to finalize the Thai's field training. We were to assist them in obtaining equipment, provide orientation on what to expect in Vietnam, then accompany them to Vietnam. The first Thai unit was made up of all volunteers from the Royal Thai Army and was called the Queens Cobras.

Incentive for them to volunteer came about through an increase in the Thai pay. Each Thai volunteer's pay was augmented so they receive the same pay as an American soldier of equivalent rank.

The initial unit of Thais contained all the service and combat elements of a U.S. infantry division, although they were only the equivalent in size to a U.S. brigade. They later became a division shortly after I rotated to the States.

When the advisory team arrived in Thailand, we were escorted by the U.S. Green Beret team to the field camp on the Laotian border where they had tents and cots set up for us. It was after dark when I went into my tent, luckily, I had a flashlight which I happened to shine on my cot and there lay a snake called the "one step," which is a bamboo viper. I pulled my 45 and shot the hell out of my bed killing the snake and creating some excitement for the other team members. The reason for the name "one step" became quite obvious if you were bitten by one. I don't know yet whether that snake chose my bed because I had been a Green Beret, or just by chance, but we will never know for sure.

In addition to a lot of hard work in Thailand the team had a lot of fun. When we were not in the boondocks we stayed in a hotel in Bangkok and attended several VIP parties. We had an audience with the King of Siam who was born in the United States and was fond of American jazz music. He had a lovely wife, the Queen, who had been educated in France. While visiting the palace we were shown the Emerald Buddha which was the King's own. The Buddha was actually carved out of jade and stood about 2 feet high. The King changed the clothing on the Buddha when he wanted to change the seasons for the country. One set of clothing was used for the monsoon or planting season, another set for the harvest season.

We were invited to two birthday parties on the same day for the Prime Minister. One in the morning until lunch for public officials, the other in the evening for his family and the military. The Prime Minister's son was the liaison officer between the Thai volunteers in our division. He was a very capable major who had graduated from West Point and had attended the Infantry Officer Advanced Course at Fort Benning. Many of the volunteer officers were graduates of foreign military schools such as Amherst, Saint-Cyr, and West Point.

I would have been delighted to have spent the rest of my tour in Thailand. It was customary for many rich Thai men to have two or three concubines, and if they had a distinguished visitor stay overnight they would provide a concubine for a bed partner. I was offered a concubine which I refused, so they gave me a car to drive. Think what you will—but remember, I had just been on R&R.

The car furnished me in Thailand

Just before the Thais and our team were to return to Vietnam, the Thais hosted a big dinner party where the bridge crossed the River Kwai; remember the movie? This was a railroad bridge linking Thailand and Burma built by British prisoners captured by the Japanese in World War II. These prisoners must have suffered terribly. There was a cemetery nearby where thousands of British soldiers had been buried.

There was a restaurant on each side of the Thai access to the bridge, where long outdoor tables were set up to feed a going away party attended by all the big Thai Army brass. I sat next to the Commander of the Thai Forces and when they passed a little sauce for our salad he nudged me and whispered, "Hot." Old smart aleck me, I had eaten hot Mexican food and thought, "Hell, this can't be too hot," how wrong I was. One bite and my mouth was full of burning needles and it burned all the way down. I grabbed a pitcher of beer and gulped most of it down, then headed for the bushes where that stuff burned all the way up again. I had to close my mouth to keep from vomiting up my toenails. I don't know what it was but I know I never want any more of it. This was the case of an advisor not accepting advice!

While in Thailand I had the opportunity to watch the elephants working in the rain forest. These magnificent animals could pick up huge teak logs and precisely balance them on their tusks. They would then carry them through the dense jungle growth. I was told these animals were captured or taken from their mother at a very early age. The baby elephants were chained to a stake or tree and left for several days without food or water. After being left alone for a time, when approached by their owner (handler) with food and water the young elephant automatically accepted the handler as its surrogate mother. These elephants were highly revered by their owners and were treated extremely well, almost like a member of their own family.

Later back in Vietnam I was asked by my Division Commander if I had any elephant handlers assigned in the Thai unit which I was advising. A herd of wild elephants had been located near our base camp and the General would like me to capture a baby elephant to be the Division mascot. I asked for some Thai volunteers who had trained elephants and set out to do the General's bidding.

We flew in helicopters over the wild elephant herd, dropping concussion grenades to drive them into an area where we could surround them with Army personnel carriers. Once we had them surrounded with the APCs we would try to separate one of the baby elephants from the herd and let the rest go. As we were driving the elephants and hovering over them in the choppers, some of the bull elephants would stand up on their hind legs with their trunks in the air and trumpet as if they were trying to chase away the things that were bothering them. I never saw a baby elephant stray more than three feet away from a grown elephant.

We finally herded the elephants near the APCs, but having observed the power of these animals I was afraid to try and surround them for fear that they could upset an APC causing damage that would be hard to explain. I radioed Division Headquarters and advised them the only way I thought we could capture a baby elephant would be to shoot the mother and this I did not want to do. Before my Division arrived in Vietnam it had been the practice to shoot all wild elephants discovered in the jungles as the VC used elephants to transport their supplies.

The General acceded to my desires not to shoot an elephant and this ended the operation to obtain an elephant mascot. The Division later captured a 12-foot Boa Constrictor snake as a mascot and a man was assigned as the snake keeper. His duties were to feed the snake and allow it to crawl through the grass to keep its skin moist.

CHAPTER 36

Thai Operations in Nam

ALL TOO soon we had to return to the hell of Vietnam where the Thais went directly into the field. Instead of setting up a base camp as the U.S. forces had done, they stayed mobile moving their headquarters in conjunction with their tactics. The Thais were excellent soldiers with an Oriental philosophy. They believe that in death all parts of the body should be interred or cremated together. This custom almost caused an international incident. The Thais were constantly encountering booby-traps and mines set on the outskirts of the Vietnamese village. When a couple of the Thais were killed from a mine, the Thai commander went into the village, lined up all the head men and told them to produce the Vietcong that was setting the mines what they would "Tạd h⁻ạw k⁻hxng phwk k⁻heā" or cut off their heads. When the head men pointed out two Vietcong, the Thai soldiers cut their heads off and tossed them down a dirt road to be mounted on stakes as a warning. Some U.S. reporter heard of this and reported that the U.S. forces were using Vietcong heads for bowling balls. To my knowledge no American soldiers were even near the area, but since the Thais wore the same uniforms as the Americans, the writer assumed they were American soldiers. An investigator was sent to the Thai camp to find out what had happened, but the Thai Commander refused to speak to him, since he was not under the command of the U.S. Forces he did not have to explain anything to him. The only comment from the Thai Commander was, "What about the five U.S. soldiers my unit found a few days earlier that had been hacked to pieces by the Vietcong?"

The Lieutenant Colonel who had been the chief advisor rotated to the States and I was assigned as the chief advisor. I was already past due to rotate since I had arrived in Vietnam with the advanced party, but it seemed this fact had been overlooked at Division since I was away and on my own with the Thais in the field. A few days before my replacement arrived, I accompanied a Thai company assisting a Vietnamese company on a mission near the village of which I spoke earlier. The Thais were on a joint mission with a Vietnamese company to recover some bodies of Vietnamese soldiers that had been killed the day before. I was riding on top of an armored personnel carrier when it hit a mine and blew me off. It then caught fire burning 12 Thai soldiers to death. The infantry companies were in a mined area and had detonated another mine causing six casualties. Without realizing I was seriously hurt, I called for an evacuation helicopter and a mine detector. When the chopper arrived we had cleared the people from the mine field and they were continuing on their mission. I used a poncho to wrap the stump of the leg of one of the Thai soldiers which had been blown off in the mine explosion. I squeezed the stump of the man's leg with my hands to keep him from bleeding to death all the way to the hospital. I thought the helicopter would never arrive at the hospital, it seemed like a lifetime. I could feel the man's life flowing out in my hands.

I had landed on my head when the APC hit the mine and the only thing I noticed of any injuries to myself was a small cut on my head. There was much more to this action which was related in the citation awarding me the "Silver Star," the third highest award given for valor.

Before I leave Vietnam I feel a need to make some comments about things I observed or became aware of. There were many, many restrictions placed on military operations in Vietnam that the majority of Americans were unaware of. One of these restrictions being that when our forces were receiving enemy fire from certain restricted areas we were not permitted to return fire until we had obtained permission on the highest headquarters in Saigon. Can you imagine being shot at and not being allowed to fire back at your enemy? I was told this came about from negotiated policies between the U.S. civilian policymakers in Saigon and the very unstable Vietnamese head of government. Many readers are probably unaware of the fact that a French tire manufacturing company owned most of the natural rubber producing trees in Vietnam. As a matter of fact, at onetime Vietnam was a French colony. Before the French were defeated and Vietnam became an independent country, the French landlords had built many magnificent homes on their Vietnam plantations. These French landlords consistently treated the Vietnamese peasants as slaves. My reason for relating this history is to acquaint you with the fact that when our forces were operating in some of these rubber plantations they were not allowed to cut a single rubber tree, even if we needed to clear fields of fire for effective use in firing our weapons or build revetments for our protection.

Our forces had to police all battle areas with great care, we were not allowed to leave any trash of any kind outside our secure base camp. The Vietcong were very resourceful in gathering and using anything we discarded.

The VC would salvage the explosive material from bombs or artillery shells that failed to detonate. This was a dangerous operation for them but it provided them with a source of explosive supplies. From the material obtained from our duds they would fashion crude but effective mines and booby-traps.

For hand grenades the VC would melt any scrap metal they could find, then place the metal in sand molds to create a metal blob with a hollow center. They would fill the center with explosive powder and place a detonator in the center. To this they would attach a hollow piece of bamboo with a rolled-up piece of paper filled with powder to act as a delay train. This delay train there was a friction cap attached to a piece of string. The VC would tie the string to their finger and when they threw the grenade the string activated the friction cap igniting the powder delay train, burning through the paper to ignite the detonator. Sometimes the powder train would be broken and the grenade would fail to go off. Often times the powder train was too short and the thrower would become the victim. I gathered several of these grenades, disarmed them, and gave them away as souvenirs. They were crude, but many times very effective. From these two examples you can understand the need to police the battle areas. While I was the Brigade Staff Supply Officer, the policing of the battle area came under my jurisdiction, and if you remember I was an outstanding police officer.

One could never be sure who was or was not a VC. I was told of little girls offering our troops baskets of fruit with explosives hidden in them. The average American soldier is a happy, sympathetic and trusting person. At first when we entered Vietnam we were very naïve and succumbed to these tactics. Later the combat troops became very hardened and towards ALL Vietnamese as we could not tell who was a friend or a foe.

It was said we were trying to help the Vietnamese become a democracy, but anyone knows that a person without legs can hardly learn to walk. Our political action people seldom left Saigon where most of the educated Vietnamese people lived and stayed. Had they visited some of the hill tribes (Montagnards), the political advocates of exporting democracy might have realized they were doomed to failure. Most of the nomadic tribes were savages that lived off the land and used bamboo bows and arrows to obtain food. We even found bamboo arrows in some of our helicopters. With all their crude tactics, the Vietcong were able to survive our political and military efforts which eventually forced us to withdraw.

Two days after the action in the mined area, I went to Bien Hoa and asked to be rotated to the States. Which I was allowed to do, although I had not received orders from Division I was told at Bien Hoa I could take 30 days leave in the States and orders would catch up with me.

I was flown into San Francisco with nothing to wear but jungle fatigues, as all officers had not been allowed to take their uniforms to Vietnam. All of mine were with Lois in Tennessee. I encountered my first war protesters at the San Francisco Airport, a dirty unkept motley mob tried to spit on me and shouted, "Baby Killer" at me. I tried to ignore them but after the experiences I had just been through, it was very hard to keep my composure. I had been killing people that had not hurt me half as much is those protesters did. I was lucky to catch a flight to Tennessee almost immediately which allowed me to get away from that dirty mob. For the first time in my life I was ashamed of the people in my own country.

On my arrival in Knoxville, Tennessee, the latter part of January, it was really cold and I was still wearing jungle fatigues. Lois met me and before we got to a motel near the airport I thought I would freeze to death. Just a few hours before, I had been in 110° heat where the jungle fatigues I was wearing were too heavy for comfort. When I arrived in Knoxville the temperature was 10° above zero and I was still wearing jungle fatigues. This amounts to 100° temperature change in a few hours. After a night in Knoxville we drove to our home, about 70 miles away, where I stayed for 30 days on leave and became reacquainted with my family.

CHAPTER 37

Fort Campbell

WHILE ON leave I started experiencing severe pain in my neck and back and was losing the feeling in my left arm. I thought I had pleurisy that was causing my neck pain, and jungle rot on my arm that caused the loss of feeling. As long as I remained immobile the pain didn't seem to bother me. So I sat in a chair with a heating pad on the back of my neck most of the 30 days I was on leave. I enjoyed being with Lois, Skeet, and Tess who had just turned three years of age. Tess had just started to recognize me as her father. About the middle of my 30-day leave I received my orders and a Silver Star Medal in the mail. I was to report to Fort Campbell at the end of my leave where I would be given on post housing for my family.

When I reported to headquarters at Fort Campbell, the General told me he was giving me command of a Battalion. This was one assignment I was eager to get. Because of the zero efficiency report I had received in Munich, my name had not been on the Lieutenant Colonel promotion list. Command slots were at a premium and the General told me the Lieutenant Colonel commanding the Battalion needed another month to complete a year in command. He assigned me as the Assistant Deputy Commander, a do-nothing position. I was told to have a set of chain of command pictures made for the Battalion I was scheduled to command. I should then take the time to turn in my medical records and get my family settled in our quarters.

A veteran of Korea and Vietnam, this General was one of the most considerate men I ever met. The Deputy Commander a combat infantry veteran, a full Colonel was waiting to retire. This Colonel had received a low efficiency report earlier in his career and he advised me to write a letter to the Department of Army explaining my situation. He even volunteered to help with the letter after he reviewed my record. I was to be given a Battalion to command in spite of not being selected for Lieutenant Colonel. I had my picture taken and went to the hospital to turn in my medical records where I asked to see a doctor. I showed him the jungle rot on my arm and told him I was losing the use of the arm. I then told him about the pain in my neck. He said he could cure the jungle rot very easily, but that wasn't what was causing the loss of feeling in my arm. He told me to go to X-ray and had my neck x-rayed and wait there for the wet x-ray reading. Then I was to take the x-rays to the orthopedic clinic where he would arrange a consultation for me.

I went to X-ray where they placed me on the table, not to gently, and took four or five pictures. They told me to wait in the hall until the radiologist could read them. I waited 15 or 20 minutes when a doctor came out and said, "Major Walker, we want to take some more x-rays." I asked him, "What's the matter, couldn't you find anything?" He replied, "We found something we don't believe." This time the technicians were very gentle when they placed me on the x-ray table to take several more pictures. When they completed the x-rays, they put me on a gurney and wheeled me into the radiologist's office. He asked me if I knew I had a broken neck. This was caused when I had landed on my head after being blown off the personnel carrier in Nam. I had not realized I was hurt except for the cut on my head (which later was diagnosed as a concussion). I had reacted to the situation instinctively, due to my training. My adrenaline was flowing, I was scared, I was concerned with all the other casualties and never even thought of myself as being injured.

When the radiologist had confirmed my neck was broken, I was wheeled down to the orthopedic ward where they put me in a traction bed. This was on Friday, when the orthopedic doctor came into the ward and told me that they were going to evacuate me to Walter Reed Hospital in Washington, DC on Monday. I asked the doctor to let me spend the weekend at home with my family. At first he said no, but then I told him it had been over 40 days since I had received the injury and that I would remain very immobile. He hesitated, but when I added I had just returned from Nam he relented. He gave me stern orders not to move around and to be back in the hospital early Monday morning for my flight. The doctor said one of the spines on my vertebrae was just hanging by the nerves and if they were severed I would be permanently paralyzed. I notified my CO what had happened, then went home and told Lois. We spent the weekend making arrangements for me to go to Walter Reed. We decided she should stay at Fort Campbell where she could be with the kids, as Skeet was in high school and Tess hadn't started kindergarten yet. We had some good neighbors on the post at Campbell and it was here that I came to realize what a saying I had heard before really meant. The saying being: "The Army takes care of its own."

For the next two years I was reminded of that saying many times in many ways. I felt I really was one of the "Army's own."

On Monday morning I reported back to the hospital where they wanted to put me on a stretcher, but I talked them out of that since I hurt worse lying down than sitting up. I arrived at Walter Reed Monday evening carrying my orders in a briefcase, a clerk read them and told me what ward to go to. When I entered the ward the nurse in charge wanted to know what I wanted and when I told her who I was she said I had been expected, but that I was supposed to have been on a gurney and not walking. The fact that I had walked in seemed to upset her, but she helped me to bed very gently and was very friendly and nice to me.

The nurse, a captain, put me in a bed near a young Lieutenant who had been in one of the First Brigade's combat units. He had been shot through the wrist and the doctors had performed several tendon grafts, but they would not heal properly. He had been at Walter Reed several months and when his wife came in to visit him that evening he introduced her and told me they had been married in Oak Ridge, Tennessee. Whereupon, I started asking about people she knew there. As it turned out, I knew her father and mother quite well, as Lois had worked with her father. It's a very small world. Her husband the Lieutenant had attended UT and had a degree in education, where he planned to teach when he was discharged. I learned later he had been promoted to Captain, discharged and was teaching in Hawaii.

Recall

Last official picture, 1969

CHAPTER 38

Hospitalization and Recuperation

THE NEXT morning I was transferred to a private room and met a neurosurgeon who was to perform an operation on my neck. After we talked for a while I told him what had happened, he told the nurse I was a very hyper person and I was permitted to do or go wherever I felt I could. The doctor was going to do a myelogram (a diagnostic procedure) on me the next day.

That night I went to the Officers Club. The doctor and my nurse were there having a drink, they asked me to join them which I did and stayed through many rounds of drinks. I kept watching how much the doctor drank because I wanted a sober doctor cutting on me.

The next morning, I was bathed, shaved, and painted with disinfectants. I was given some pills to sedate me, as I was to remain awake during the procedure. I was then placed on a gurney and wheeled to an operating room that contained an x-ray machine. Dye was injected into my spinal canal, then I was x-rayed from my butt to my head. I was turned and tilted in every direction which cost me much pain, and when it was over I was told I must lie flat without moving for 24 hours. They had withdrawn most of the dye, but I was told that if any dye remaining entered my brain it would cause a terrible headache.

I talked to Lois by phone every day, except two, while I was in the hospital. The General at Fort Campbell made arrangements for a "hot line" into our quarters where she could call me at any time. A team of neurosurgeons did the operation on Friday. A piece of soft bone was taken from my hip and placed in my spine at the neck. A platinum clip was placed in my neck to hold the spine in place. I have often told people if the platinum was removed from my spine, the gold from my teeth, the iron from my blood, and the lead from my rear end, I would be worth a small fortune.

The doctors at Walter Reed confirmed that the vertebrae was just hanging by nerves, and if the nerves had been severed I would have been completely paralyzed on at least one side of my body. I was kept in intensive care that night and Saturday. On Sunday I was taken to a ward and a metal brace put on my neck. I was to be taken to the cast room the next day. That night the man in my bed next to me died which really upset me. So when they took me to put the cast on, my doctor made arrangements to have me brought back to a private room.

I was placed in a plaster cast from the top of my head to my waist. Only my eyes, nose, and mouth showed. They even cut small holes for my ears so I could hear. This was a heavy bulky thing called a Minerva Vest.

My doctor and I became good friends, so I began asking him when I could return to Campbell. He finally told me he would release me when I was able to walk to the transportation office and get them to cut orders. I did just that on Thursday and managed to talk to a Sergeant into cutting flight orders on a military evacuation plane for the next day. I showed the orders to the doctor and he laughed, but he signed my release and call the Fort Campbell Hospital to assign me a room on arrival.

I boarded an Air Force plane that flew me over half the United States, and on the third landing I asked one of the nurses aboard when we were going to arrive at Fort Campbell. She told me the plane was not going to Fort Campbell, but to Wright-Patterson Air Base in Ohio where they would put me in the hospital until another plane could fly me to Campbell. I was probably responsible for this screw up by talking the Sergeant into getting me on the first available flight. Another case of doing things "my way" instead of the "Army way."

The next stop was at Johnson City in East Tennessee, and when we landed I draped my uniform coat over the cast and starting to get off the plane. The nurse, a Captain, began telling me I could not get off the plane. I pointed to the Major leaves on my coat and she walked away. She started crying and called the pilot who asked me what I was going to do. I told him I was in Tennessee and I would call the Highway Patrol to take me to Campbell if I had to, because Tennessee was my home. I then walked off the plane.

Again my luck held, as there was a civilian plane leaving for Nashville in 20 minutes with one seat open. I just had time to pay for my ticket and call Lois to meet me. I looked like hell in that bulky cast with my uniform coat draped over it. When we landed in Nashville two nice people offered to assist me to the Veterans Hospital where they thought I was going. I thanked them and told them my wife was meeting me. When Tess saw me, she was scared of me and started to cry. This tore my emotions apart. I spent 10 weeks in the cast assigned to the hospital, but I was allowed to stay in our quarters on post.

Tess would not come near me until we told her I was wearing a "Munster" suit, as *The Munsters* was a TV show she watched at that time. She began peeking around the bedroom door and finally with her mother's help she came in to scribble her name on my cast. After that I would see all kinds of little people peeking around my bedroom door. Tess would invite all her little friends in the neighborhood to come see her daddy's Munster suit. My cast became the drawing board for the neighborhood kids.

After the cast was removed I was given three months convalescent leave and when Skeet graduated from high school we spent most of the time at our new home in Tennessee.

When I reported back for duty the Battalion Command had been filled, so the General wanted to assign me as the Inspector General as the present IG was retiring. The IG had to be approved by the IG at the Department of Army and I told the General I did not think I could be approved, but he sent my appointment in anyway.

He told the incumbent IG to write a letter on my behalf explaining the occurrence in Munich. After my career branch received the General's letter, the Department of Army requested me to send the details of the occurrence at Munich. The Deputy Commander helped me write the letter and put a very strong endorsement on it. I sent the letter with very little hope that it would do any good. In less than a month I was promoted to Lieutenant Colonel and appointed IG.

The Department of Army sent me a letter saying they had reviewed my records and the zero rating was so far out of line with all my other efficiency reports that it would be removed from my permanent file. They had then looked at the efficiency reports that Beebo had given the other officers and found them all to be very low. I think the Lieutenant Colonel who endorsed the report retired and Beebo was given a letter of reprimand and released from the service. The Department of Army had backdated my promotion to the date I should had been promoted, which was about six months earlier, but they did not give me the difference in pay.

I served as the IG at Fort Campbell until I started having pain in my lower back and was sent back to Walter Reed for three more operations on my spine. While I was serving as Inspector General, the following quote was handed me:

> "The typical IG is a man past middle age, cold, passive, non-committal, with eyes like a codfish, polite in contact, but at the same time unresponsive, calm, and damnably composed as a concrete post or a plaster of Paris cast; a human petrification with a heart of feldspar and without charm or the friendly germ; minus bowels, passion or a sense of humor. Happily, they never reproduce, and all of them finally go to hell." -Anonymous

The last time I was at Walter Reed it was determined that I should undergo a medical board evaluation. I was examined by specialists from every field of medicine and each issued a condition report. The neurosurgical report said the major nerves in my legs had been injured and were degenerating.

The report said I would probably be in a wheel chair within a year unless I could stand the pain and keep exercising. That has been almost 30 years ago and I'm not in a wheelchair yet. When the psychiatrist made his report, he wrote, "This patient suffers from extreme combat stress syndrome." I asked him what the hell that meant. He told me, with a smile, "That's the nicest way I know to tell you you're crazy."

To give readers a second opinion I wish to quote a paper Tess submitted in college, while working for her Masters degree in Psychology:

"I believe that adjustment means to be able to adapt to surroundings. By this I feel that a person should be subject to change within society and other surrounding elements of the environment. The best well-adjusted person that I know is my father, Lieutenant Colonel Arthur Walker. He is a stable and proud man. He ran away from home at the young age of 12 with nothing but the clothes on his back and a few dollars in his pocket. He did not realize that within just a few short years that he would enlist in the Army and fight in the next three wars that would change his life forever.

The worst of the wars was of course Vietnam. During this time Arthur was a Green Beret and would be promoted to the rank of Lieutenant Colonel. While in combat he was blown up which his neck and back was broken. While in the hospital, doctors told him he would never walk again. The doctors were wrong or at least my father proved them to be. To this day he is walking in a perfectly normal way

Lieutenant Colonel Walker has seen friends die in his arms and one of his regiments of which he was in command shot to hell. To imagine all of this in my mind, would be impossible. I know my father remembers his trauma, but he has adjusted and is a calm, happy person.

Today he is retired and has made a respected name for himself in our community. He now makes wooden toys in his workshop and he loves deep-sea fishing. Speaking of fishing, he and my mother have taken our boat to Florida for two months of fishing and sun; the only complaint I have is that they left me here to battle the snow, ice, and classes."

A note from the instructor said: *"Perhaps your father would like to read this."*

During the year and a half I was having these medical problems, I was carried on orders as the post IG. This gave me an assignment for rating purposes which would help my record if I were returned to full duty status. In truth, I did very little except to act as counsel to two majors assigned to my section who did the real work. For almost two years I was undergoing an operation or was on recovery leave.

My daughter Tess while in college when she wrote

my second opinion:

The Best Adjusted Person I Know is My Father

One episode that was called to my attention while I was the Inspector General at Fort Campbell was a great concern to a mother who wrote me a letter.

It seems that a young soldier stationed at Fort Campbell for basic training at impregnated a young girl while home on leave. I received a very serious letter from the mother of the pregnant girl.

The letter was handwritten so I will attempt to paraphrase the letter as I received it"

Dear Sir, one of your soldiers was home on leave and got my daughter "knocked up," and I want to know what <u>you</u> are going to do about it? I have written to his company and they say they don't have a soldier with that name. (Soldier's name omitted.)

Some people told me to write the Inspector General as he is the "big Cheese" in this matter and if you <u>ain't</u> the big cheese then tell me who is? (Signed name omitted.")

Although this letter might appear humorous to some people it was a very serious matter to that mother and was considered seriously by me. I had one of the majors assigned to the section investigate the matter.

The Major reported the man in question had received orders and had departed Fort Campbell a month earlier. We then forwarded the matter to the Inspector General in Washington, DC. I never learned how this matter was concluded, but I am sure it underwent serious investigation by the <u>Big</u>, <u>Big</u> Cheese.

CHAPTER 39

Final Review

SKEET HAD started college at UT and had completed two quarters where he was "good timing" instead of studying. He had met a nice girl at UT and they were both studying veterinarian medicine. They had fallen in love and he told his mother and I that they were going to get married. When I asked him how he planned to support a wife, he said he would quit school and join the Army. I told him he had seen only one side of military life and that was the very best side.

He had grown up in the Army as an officer's son where he had been given many privileges, but he was determined to join the Army in spite of my advice. He went to Nashville to enlist. I was called and asked if I would administer the oath of enlistment, which I reluctantly did.

He was stationed at Fort Campbell where he took basic training, but he never told anyone in his company that his father was the IG. He would call his mother to do his laundry, but he would always meet her away from his company area.

He was chosen from his company twice to be the General's orderly for a day. This was a program that was rotated among the companies to give recognition to outstanding trainees. The second time he was selected, the General saw his name tag and ask him if he knew Colonel Walker and I think that was the only time he ever told anyone I was his father.

After the medical board had convened I was back at Walter Reed, and Maury who was stationed at the Pentagon sent two officers to tell me I did not have to accept the medical board's finding. I could be kept on active duty assigned to research and development, where I could do any assignments I felt able to undertake. Because of my excellent combat record I could be retained on active duty until I had completed 30 years.

There were two reasons why this was not to my liking. The first was that I would be stationed in Washington, DC. The second was that I could not be promoted because I would not be able to pass a final type physical, so I elected to accept the finding of the medical board.

The board recommended I be retired with an 80% disability rating and placed in the Temporary Retired Reserve, which meant I had to take a physical exam every year. If I recovered I would be called back to active duty. I phoned the chairman of the medical board and asked him why I was being placed in the Temporary Retired Reserves, instead of the Regular Retired Reserves. He told me to hold and he would ask the head physician about it, and when he returned to the phone, he said, "The doctors agree that your disability is permanent and you will be placed in the Regular Retired Reserves." I was told they had recommended the temporary list for my benefit, so that if I was called back to active duty I could be given a two-grade promotion which meant I would be a Brigadier General.

So, on my 44th birthday, I was retired at Fort Campbell and a parade was held in my honor. The same day I retired my son completed basic training and was standing in the ranks. The General knew this and had Skeet brought up on the reviewing stand where he, a basic trainee, stood beside the General and myself to take the review. The Army gained much publicity from this review. Our pictures were taken for the *Army Times* and all the local newspapers. Several reporters interviewed both Skeet and myself. They did one long item entitled: *"Like Father: Like Son! Well, Maybe, Not Certain."*

In this article, I was asked whether I had influenced Skeet to make the military a career. I stated I had not and I felt he was old enough to make his own decisions. He had been accepted for OCS (Officers Candidate School) and I did tell him that if he was to stay in the Army it would be very much to his benefit being an officer.

Skeet went to OCS and had almost completed the course when he cut his thumb on a saw while making a snake cage. They were going to recycle him, which meant he would be required to repeat the course, this he elected not to do. He was promoted to Sergeant and served his four-year enlistment. He then got out of the service and he and his wife went back to UT where they both graduated.

This retirement review was the end of my military service. It was a sad occasion for me because I had served 22 years with some of the finest people America has to offer.

My son Skeet in Basic Training at Fort Campbell, KY

Facing the unknowns of civilian life, not knowing if I would be able to continue walking, not able to get a job, a wife to support and a daughter just starting school was the most traumatic period of my life. I had come from a closed military environment where I was well known. Remember the saying, "The Army takes care of its own." In a sense the Army was my family where I had found a home. Now I was thrust among civilians who were strangers to me. It seems there is much more competition among civilians than I had ever encountered in the military.

The war was still going on in Vietnam and I was not given the same welcome I received after World War II, but just the opposite. Some places in America, I felt I was regarded as a worst enemy in the Vietcong. Except for my home community in Tennessee, I was reluctant to mention I had been in the military.

There were very few demands for my skills as a rifleman or machine gunner. I could have used a pistol or taken up assassination for hire, but this would have been criminal and if caught it would have ruined my life. I could not have done this under any circumstances even though I had received training in this line of work.

The second time I was in Walter Reed, I tried to talk the doctors into sending me back to Vietnam from where I would not have returned. They said this could not be done, as a man in my condition would be nothing but a big liability to the people serving there.

I was referred to a good psychiatrist, who urged me to talk about my experiences which is one reason for this book. It has taken 26 years for me to start writing about my experiences. I started to record them on tape many times, but it never sounded like the truth. Even now much of this seems to be a fantasy of my mind.

CHAPTER 40

Conclusion

DURING MY retirement years I have worked very hard to heal my body. I have trained bird dogs walking over rough terrain, vomiting from the pain for yards at a time, but I always made myself keep on walking. I started deep-sea fishing as therapy and got so good at it that Lois insisted on my obtaining a charter captain's license. I really did this to prove to my father-in-law that I was the "best" fisherman he ever knew. I even made him become a real fishing buddy.

I have led a rather mundane life trying to do many things to occupy my mind with new thoughts. The wounds to my body have healed or gone into remission and this book is an attempt to heal my mind or at least put the memories into remission.

Some final thoughts on the military operations in Vietnam. First, in my opinion there was a violation of the most basic of tactical rules called "Piece Meal Commitment," where a smaller force than needed to accomplish a mission is committed, then another small force is committed to bolster the first force and so on. The basic rule is always commit a superior force to overpower a weaker obstacle.

A second tactical violation was the program of establishing strong point defensive bases, out of which search and destroy excursions were deployed, then returned, leaving the enemy to reenter and again control the area. This was essentially the same tactics the French used at Dien Bie Phu where they were soundly defeated. Our tacticians should have learned from the mistake made by the French. This tactical mistake occurred in Vietnam and was written about in all military history books.

It's a well-recognized fact that if you take ground you must occupy it, which we did not do. When our forces fanned out into an area the Vietcong moved out or went underground. As soon as we pulled back into our base camps the Cong reoccupied the area as if we had never been there. Population control is a must in warfare.

In my opinion, the biggest blunder came from not having a firm objective or plan in the beginning. In WWII, the objective was unconditional surrender of all enemy forces which was accomplished. With this firm commitment we won the war.

In Korea, the "quasi" objective was the 38th parallel which was negotiated, which essentially led to what might be called a draw, based on a political decision. This amounted to a conflict between the military advisors and political advisors that was never resolved.

I have just finished reading the former Secretary of Defense's book, *In Retrospect*. In it according to my interpretation, he admits to some of the mistakes made in Vietnam. There was never a positive objective established. There seemed to be wholesale violations of military tactics. No one seemed to take charge or dominate in making decisions.

There was little regard for military advice. There seemed to be a conflict in estimating the capabilities of the Air Force. Saturation bombing is a great deterrent but until the ground is physically occupied, and the population is controlled by ground forces, no conflict can ever be successfully concluded.

I could go on and on citing mistakes that were made as many others have done, but that is not the purpose of my book. I will conclude by saying, because of these mistakes, we lost our first war ever.

In summary, the United States won in World War II, tied in Korea, and lost in Vietnam.

The preceding platitudes are easy for me and others to expound on, it's like being a Monday morning quarterback in a football game, except war (or a police action) if it's to be considered a game is certainly a very deadly one. The term "police action" is an abhorrent term politicians came up with to absolve themselves from blame when they failed to pay the military their just dues. A person is just as dead if he or she died in a police action or a war, it's all a matter of semantics.

Since retiring I have been able to watch the military actions (I don't know what politicians call them) on TV in Panama, Grenada, Desert Storm, and Haiti as a civilian would view them. I get an indication that the major tactical mistakes made in Vietnam were not repeated, so maybe our Commanders in Chiefs (there have been three) are getting and listening to some sound tactical advice.

The United States is now being confronted with the problem in Bosnia and it is my fervent hope that if our forces are committed there, it is done with a resolve to win at all cost and not create another Vietnam.

After 25 years, America is still paying for the political division created in Vietnam. There is talk of time healing this rift, but the destruction created in our military establishment will not be healed until all the Vietnam generation and their children have passed through this world. Think about the time it has taken to unite Americans after the Civil War. Over 100 years have passed from the time the last Confederate and Union soldiers confronted each other on the field of battle.

In conclusion, I have been a survivor for 72 years of American citizenship. I have seen good times and bad times. I have suffered pain and sorrow, as well as good feelings and happiness. I have seen peace and war. I have seen my country united and divided in the political arena. I have been in many countries of the world where I have observed many different forms of government. I have witnessed poverty, starvation, and greed as well as wealth, influence, and philanthropy. I have been subjected to ugliness and beauty and have survived.

I now raise orchids as a hobby; they are so beautiful after all the ugliness I have seen. I will now go to admire and talk to them and if they talk back I'll know that my psychiatrist in Walter Reed was not kidding.

About The Author

In Loving Memory:
Lt. Col. Arthur R. Walker, US Army, Ret.
Sept. 22, 1923 - Feb. 2, 2017

LT. COL. Arthur R. Walker enlisted in the Army on November 11, 1943 at the age of 20.

He became a member of General Mark Clark's U. S. Fifth Army and participated in the P.O. Valley, Rome Arno River and Northern Apennine Mountains campaigns. In 1944 the U.S. Fifth Army along with Lt. Gen. Oliver Leese's British

Eighth Army ended stalemates on the Gustav Line, advanced up the Liri Valley, captured Rome and pursued retreating Axis forces across the Arno river into the northern Apennine Mountains.

In addition to the Italy campaigns, Lt. Col. Walker participated in North Africa and Middle Eastern Europe campaigns and was awarded the EAME (European-Africa-Middle Eastern Campaign Theater Ribbon) along with three Bronze Stars. He also received the Good Conduct Ribbon and WWII Victory Medal.

On November 20, 1945 due to demobilization of his unit, he was transferred to the U. S. Army Reserves.

On April 18, 1952, Lt. Col. Walker was recalled to U. S. Army active duty and served until September 26, 1969. During that time, he served in Special Forces and the Green Beret.

Education and Training

- Infantry School
- Army Infantry Command Course
- Airborne School
- Special Forces Staff Officer School
- Counterinsurgency and Special Warfare Officers Course
- Inspector General Orientation Course
- Burmese Language
- Code of Conduct
- Course A Military Justice Training
- Course B Military Non-Judicial Punishment
- CBR (Chemical, Biological, Radiological and Nuclear Attack) Training
- Geneva Convention

Decorations and Awards

- Republic of Korea Presidential Unit Citation
- United Nations Service Medal
- Parachute Badge
- Armed Forces Reserve Medal
- National Defense Service Medal with Oak Leaf Cluster
- Vietnam Service Medal
- Combat Infantry Badge
- Distinguished Flying Cross
- Air Medal with 3 Oak Leaf Clusters (Meritorious Achievement while participating in aerial flight)
- Vietnam Campaign with 60 device and 3 Bronze Campaign Stars
- Bronze Star Medal with Oak Leaf Cluster (Heroic achievement or service in combat zone)
- Silver Star Medal (3rd highest decoration for valor in combat)
- Meritorious Unit Commendation Medal
- Purple Heart
- Vietnam Cross of Gallantry/Palm (Deed of valor or heroic conduct while in combat with the enemy and Palm was Armed Forces level, 1st degree)
- Army Commendation Medal with Oak Leaf Cluster
- American Defense Campaign Medal
- European Campaign Medal with 3 Bronze Stars
- Good Conduct Medal
- WWII Victory Medal
- European Occupation Medal
- Overseas Combat Bar (7)
- Korea Service Medal

Lt. Col. "Art" Walker, Ret.

In September 1969, Lt. Col. Arthur R. Walker separated from active service due to permanent disability having served over 20 years in the U. S. Army.

He was a devoted husband to his wife, Lois, of 68 years, and a devoted father of two children: Arthur R. Walker, Jr and Terry J. Walker.

He was also a friend to many who knew and loved him.

Some of Lt. Col. Walker's awards given for his five years of combat service in WWII, Korea, and Vietnam